TEACHING MATH CAN BE FUN

Mary Lou Nevin
California State University, Bakersfield

KENDALL/HUNT PUBLISHING COMPANY
4050 Westmark Drive   Dubuque, Iowa 52002

Cover image © 2002 PhotoDisc, Inc.

Copyright © 2002 by Mary Lou Nevin

ISBN 0-7872-9776-3

Kendall/Hunt Publishing Company has the exclusive rights to reproduce this work,
to prepare derivative works from this work, to publicly distribute this work,
to publicly perform this work and to publicly display this work.

All rights reserved. No part of this publication may be reproduced,
stored in a retrieval system, or transmitted, in any form or by any
means, electronic, mechanical, photocopying, recording, or otherwise,
without the prior written permission of Kendall/Hunt Publishing Company.

Printed in the United States of America
10 9 8 7 6 5 4 3 2 1

Table of Contents

Chapter 1
OVERVIEW 1
   Historical Development of Mathematics Education................................4
   Mathematics is Sequential...9
   Teaching for Understanding 14
Teachers Make the Difference..16

Chapter 2
PLANNING FOR MATHEMATICS INSTRUCTION 20
New Emphasis in Mathematics..21
Assessment Through Diagnostic Teaching 24
Piaget's Suggestion for Learning a New Concept 25
Suggested Sequence for Development of any Concept..........................27

Chapter 3
WHAT IS ASSESSMENT? 39

Chapter 4
MULTICULTURAL MATHEMATICS 45

Chapter 5
CONTROVERSY IN MATHEMATICS EDUCATION............................53

Chapter 6
PATTERNS, FUNCTIONS AND LOGIC 57
   Attributes 60

Chapter 7
GEOMETRY 76
   Quilts..82
   Tangrams...83
   Geoboard Geometry 84
   Geometric Solids...88
   Angles...88
   Area of Triangles 90
   Volume 93

Chapter 8
MEASUREMENT 97

Chapter 9
PLACE VALUE 112
Language of Place Value...117

Chapter 10
BASIC FACTS 126
Addition and Subtraction of Whole Numbers.......................................127
Whole Language and Mathematics 135
Strategies For Mastery..136
Multiplication and Division of Basic Facts 142

Chapter 11
MULTIDIGIT OPERATIONS 153
Addition and Subtraction of Multidigit Numbers 154
Multidigit Operations for Multiplication and Division 162
Division of Multidigit Numbers 166

Chapter 12

RATIONAL NUMBERS 169
Fractions 170
Relating Fractions to Decimals 190

Appendix A A-1
    Literature and Mathematics
Appendix B B-1
    Geoblock Activities
Appendix C C-1
    Apple Attribute Tree and Attribute Tree
Appendix D D-1
    Attribute Activities
Appendix E E-1
    Addition Chart
    Multiplication Chart
Appendix F F-1
    Fraction Bars
    Decimal Bars
    Percentage Bars
Appendix G G-1
    Historical Checking Strategies

Index I-1

PREFACE

This book is for teachers who are pursuing a career in elementary teaching. Although the focus is on mathematics in the elementary classroom, it is realized that mathematics cannot be taught in isolation. We must depend on the components of the language arts i.e., literature, language and writing to assure that students understand the skills and concepts introduced. In order for students to reach their potential in learning mathematics, the skills and concepts must be presented in a developmentally appropriate curriculum (not age level) where elementary students are active participants in their own learning. The book uses the guidelines set forth by the National Council of Teachers of Mathematics and the State of California.

The second purpose of this book is to help those future teachers who are "math anxious" find a way to enjoy mathematics and be able to teach it without passing their anxiety on to their students. There are many students who are pursuing a teaching degree who are math anxious (including my own daughter). If these students are presented with the skills and concepts in a manner which is non-threatening, they will be able to overcome their anxiety. These students have been heard to say they felt they could now feel comfortable teaching mathematics.

Although this book is in its first edition, it has been used for many years in an Elementary Mathematics Methods class. It has been updated and new ideas added throughout the years, and it will continue to be updated and revised. The students who have used this book over the years have come back and reported they keep it handy for reference. Their comments and encouragement have enabled me to continue to work on this book.

This book is not comprehensive. It is a beginning! There is never an ending when teaching elementary aged students.

Many of the ideas for this book came from my graduate work at Arizona State University. My professors: Drs. Engelhardt, Knaupp and Bitter helped make mathematics exciting for me, and instill in me the desire to help future teachers become excited about the teaching of mathematics. Teachers do make a difference!!

I would like to thank my family (Patti, Joni, Jeff, Ben, Jair and Dane) who kept insisting that I publish this book and supported me during the final rewrite. Also the fellow faculty member who have been using this book for the past several years and have given me feedback on needed changes or updates.

CHAPTER 1

OVERVIEW

Mathematics! A word that causes a wide variety of feelings! Some people react with excitement while others react with concern, and still others have no reaction since it is needed in everyday life. Many say they were never good at it, and research says that mothers tell their children they don't expect them to be good in math because they were not.

Why are there so many mixed feelings about mathematics?

The reasons vary, but they usually are related to experiences in either elementary or high school. Students say they had a wonderful teacher or a teacher who punished them when they made wrong answers. Those who have negative feelings tell about severe punishments such as being hit on the hands, sitting in the corner, being spanked in front of the class or a variety of punishment for wrong answers. A longitudinal survey indicates that the majority of prospective teachers are "mathematics anxious" (Nevin 1980-2002).

The study indicated that teacher effect had the major impact on students' feelings about mathematics – whether positive or negative. The study also showed that parental influence on a person's feelings about mathematics was neutral. Father figures seem to have more influence than mother figures on those who liked mathematics; however, the influence is small compared to the teacher influence. The mother figure seems to play the role in letting their children know they were never good at mathematics and did not expect them to be good at it or like it. How do children, who are consistently told this, develop a positive feeling for mathematics?

How does attitude affect the prospective teacher?

New teachers have their own opinions regarding mathematics, and their opinion may affect the way they teach mathematics. This is why they must reflect on their feelings about mathematics. The goal is to empower all their students in the area of mathematics. To do this we must have teachers who are dedicated to helping children learn.

Individuals who have not excelled at math, nor had high grades in math can still become excellent teachers of mathematics. Sometimes we find those who found math easy in school and had little difficulty learning cannot see the underlying skills needed by the children. They cannot see the small intricate steps some students must follow to learn new concepts. On the other hand those who have struggled with mathematics all their life may become more effective teachers, because they know the small steps, and have the patience needed to be successful teachers. The past does have an effect, but it is the desire of the teacher that has the greatest influence on the teaching and learning of mathematics.

Future teachers need to reflect on mathematics with an open mind. One way to help determine the effect they will have on their students is to think about the following questions:

- *What is mathematics?*
- *How should mathematics be taught?*
- *Who should determine what should be taught?*
- *Is there a reason why students are not learning mathematics?*
- *What effect will the lack of mathematical skills have on students?*

We hear these questions daily and in many forms especially when the state is trying to make changes. The State Departments of Education are being questioned about their resistance to change. Although mathematics has not changed a great deal over the decades, the vocal few make it seem as if it has changed.

Textbook adoption can be a political movement. A few years ago this happened after the state had adopted new textbooks. The state committee (teachers and lay people) spent thousands of hours reviewing textbooks to determine which ones best fit the needs of the students and met the guidelines set forth. Once their choices were announced, others came out against the adoption. This group of people forced additional textbooks to be adopted. They wanted a more traditional text – not the "fuzzy" mathematics they felt the adopted books presented. A small group of vocal people felt they had won because they had a strong influence on the state adopted math framework. Items were introduced much sooner, and many concepts seemed to have a misfit with the developmental levels of children.

Once the new standards were adopted and closely aligned with the new SAT 9 test, additional textbooks were added to the original list. Districts then had the option of adding the new textbooks to those they have already purchased.

Adopted standards were another point of controversy. Again the state committee met and developed standards, which supported good pedagogy and met the needs of the population. These standards were questioned, and new members were added to the committee, and other standards were written. Politics and standards seem to be compatible with each other. New standards were adopted, and the changes in math teaching began. The National Council of Teachers of Mathematics (NCTM) provided assistance to discuss the new standards – but to no avail. They could only react to what was being proposed, however, they could not make recommendations. The new standards are in every school within the state. Time will tell which group was correct. The major question is "Will the students in the state be able to meet the high level requirements of this new committee".

Change is difficult, and change in textbooks is even more difficult. People have been watching textbooks since the late 1800's --and the textbooks follow the same format. The only changes are in the price of materials, and updating the material. The new textbooks are more visual than the older ones. The textbooks of the late 1800's consisted of problems only on each page.

The student's ability to do mathematics (or arithmetic) has not changed. The textbooks of the late 1800's stated students did not know how to do mathematics, and something must be done to improve their mathematics skills. We are still hearing the same thing today. The difference being the textbooks in the late 1800's focused on a few students who were entering business, while today's textbooks focus on the entire student population. Time has not erased the student's lack of ability to do mathematics.

Why has this statement been unchanged for the past 100 years?

Change meets with resistance. Research has shown many teachers are afraid of mathematics, and have not been trained in ways to present mathematics in a manner that is understandable to all students. The problem with mathematics is it's always someone else's fault. The blame was placed on the "new mathematics" several decades ago; then it was on the lack of memorization of basic facts, etc., and then the "new new" mathematics. The "new new" mathematics focused on the students becoming active participants rather than passive listeners. The hope was teachers would move away from "open your books and do "Y" page", to the development of concepts and understanding for the purpose of problem solving. In order for this to occur, teachers would have to share the classroom with the students, and give some of the control to the students. This concept would require teachers to be secure in their ability to do mathematics and to explain mathematics. The demand for 'back to basics', has again given teachers the option to "teach from the book".

The atmosphere of the mathematics classroom should change from a quiet place (zero talking), where everyone is working diligently on the same page to a classroom humming with constructive noise. Students should be interacting and discussing various problems. The concern should be the process of solving the problem rather than just obtaining the correct answer --a process that was encouraged in the textbooks of the late 1800's

Another change in the classroom will be the use of assessment. Assessment will come in many forms: journals, rubrics, writing, portfolios, computations, etc. The focus on assessment will be the process of obtaining an answer – not necessarily the correct answer. Incorrect solutions will enable the teacher to diagnose the solutions and provide guidance in helping the students find the correct solution.

The major reason for the change in mathematics is to help bring mathematics up-to-date. We can no longer say what was good enough for our grandparents is good enough for us. We must look at where our students are going, and how we are going to help them become citizens of the future. Change does not mean getting rid of all the old and starting with the new. Change means intertwining the new with the old and making a program that will benefit all of the students.

HISTORICAL DEVELOPMENT OF MATHEMATICS EDUCATION

The development of mathematical ideas has taken many different directions throughout history. When we look back to the early centuries, theorists felt children learned by doing. Comenius (1592-1670) believed education should be a positive learning experience than included freedom, joy and pleasure. He also believed that education should follow the order of nature. This theory was later reflected by Montessori and Piaget. Both stressed learning through the use of the senses. Rousseau (1712-1778) agreed with Comenius. He felt we should follow the child's natural development of learning. He felt the child should develop according to their natural ability. Pestalozzi was a follower of Rousseau and followed his theories. He also added that the best teachers were those who taught the children and not the subjects. Froebel (1782-1852) [father of "kindergarten"] studied under Pestalozzi. He spent his life developing a curriculum and methodology for young children – some of which is still followed today.

We have seen some major changes occur during the late 19th century. The mental discipline theory of learning had a great impact on teachers – and is still felt today. This theory stated the mind was a muscle that must be exercised in order for arithmetic to become a part of the long-term memory. This gave the teachers the right to give students long exercises in computation in order to stimulate their minds. Edward Thorndike's stimulus response theory replaced the mental discipline theory. His theory stated there was a connection between the stimulus and response. His original emphasis was on having students use manipulatives to begin the connection but delay drill until third grade. His drill and practice theory became an easy way and the only way for many teachers to teach arithmetic. They did not need to know arithmetic in order to teach using this method; therefore, the arithmetic book became their security blanket. He stated that students should be able to do "x" facts in *x* minutes thus the birth of timed tests as a determinant to whether a student knew arithmetic.

Following this era, a number of theorists began to state there was more needed than this stimulus response theory. They felt that children needed to develop an understanding of arithmetic and be able to apply it to mathematics. John Dewey felt that children needed many experiences in mathematics and they should move from manipulatives to the written algorithm. He also believed that children should be part of their own learning. This was a break from Thorndike's drill and practice emphasis. Dewey's theory met with objection from teachers. William Brownell and Jerome Bruner also reinforced this theory. William Brownell stated in the 1930's that in order for learning to become permanent, there had to be understanding. He felt learning began at the concrete (manipulatives) level and then proceeded to the pictorial level (drawing of pictures) and finally to the abstract level (written numbers). This process helped assure understanding. Zoltan Dienes expanded on this concept and stated students needed different experiences to learn the same concept. He entitled his theory "Multiple embodiments". Although all theorists have their own terminology, they agreed there was a developmental process in learning mathematics. They stated that children needed to be actively involved with manipulatives at the first stage of development - the concrete state. This should be followed by the pictorial stage where pictures replaced the manipulatives. The final stage was the abstract stage where children were writing algorithms - the stage which many think is mathematics. This is the process which many of us used during our own education – we knew nothing of the concrete or pictorial stages. Jerome Bruner has been quoted as saying that you could teach anything to a child as long as it focused on his/her developmental level.

Jean Piaget followed and emphasized that children develop in stages. He stated that age was not the determinant of what children should know. He emphasized a developmentally appropriate curriculum that allowed children to develop at their own levels, not the levels of adults. His theory helped change the curriculum and allowed teachers to plan instruction for the individual child.

Another terminology heard in regard to teaching mathematics is the "constructivist theory". This theory closely follows Piaget's theory. It states that children should be allowed to discover their own facts, investigate mathematics and draw their own conclusions.

All of the above theories have had an influence on the teaching of mathematics. The original textbook adoptions focused on the developmental theory as well as constructivist theory. The books have been updated to follow what theorists have found to be successful in teaching children mathematics. The controversy is that books do not emphasize drill and practice. However, if you look at textbooks from the late 1800's up to the present time, the main difference is the format of the books, not the content.

As teachers move from era to era, they must determine which era of mathematics they will follow. It is hoped that they will find a combination of all theories. No one has said that the drill and practice should be set aside. A developmentally appropriate curriculum that focuses on students understanding mathematics will include a certain amount of drill and practice. As a teacher, you will be the decision-makers regarding your class - not the textbook.

The textbook is to be used as a guide to teaching. The teacher must make the decision of what the students need to know in order to learn the lesson, how the material will be presented, and what drill will be needed to emphasize the skill.

WHAT DETERMINES THE MATERIALS WE TEACH?

The National Council of Teachers of Mathematics (NCTM) *Principles and Standards for School Mathematics* and the *Mathematics Framework for California Public Schools* are the guidelines used in determining what should be taught in mathematics in California. The two documents parallel each other and give suggestions for teaching mathematics. Both documents focus on the use of manipulatives, understanding, and application of mathematics. Both of these documents were revised for the year 2000.

Textbooks also determine what is being taught in mathematics. Although each state has its own guidelines for selecting textbooks, the states of California and Texas have a major impact on textbook series. The textbooks used in the state of California must meet the requirements stated in the NCTM standards and the California Framework. Teachers, parents, and professionals review each and every book in a series to determine if it should be submitted as a possible state adoption. The committee makes a final decision that is presented to the State Department of Education. These books are placed on a state-adopted list, and schools choose from this list. The current adoptions are different from the past adoptions. The traditional textbooks were limited, and "applications, thinking, and doing" textbooks became the major adoption.

What Does the NCTM Standards say about Mathematics in the Elementary School?

Goals for Students:

Goals for students must reflect the importance of mathematical literacy. All students from Kindergarten through 12th grade should:

- become confident in their ability to do mathematics
- become mathematical problem solvers
- learn to communicate mathematically
- learn to reason mathematically.

In order to meet these goals, students must be exposed to numerous and varied mathematical experiences that will encourage them to value mathematics, to develop mathematical understanding and appreciate mathematics. Students should be encouraged to explore, guess, and make and correct errors to assure their confidence in their ability to do complex problems. They need to be able to read, write and discuss mathematics in order to build problem solving skills.

John Dewey's (1916) distinction between knowledge and the 'record of knowledge' may clarify this point. For many 'to know" means to identify the basic concepts and procedures of the discipline. For many who are not mathematicians, arithmetic operations, and algebraic manipulations, geometric terms and theorems constitute the elements of the discipline to be taught in grades K-12. This may reflect the mathematics they studied in school or college rather than clear insight into the discipline itself.

What are the Grade Expectations for Mathematics?

The NCTM stated the following are grade level expectations!

Kindergarten through Grade Three

The mathematics program that children encounter in school must build on familiar experiences to extend the children's understanding and appreciation of mathematics. The Instruction program should be:

- ✓ A Classroom atmosphere that fosters the development of logical thinking and problem solving.
- ✓ Students use concrete materials, often in varied activities.
- ✓ Students have many opportunities to explore, investigate, and discover.
- ✓ Students are continually encouraged to interact with each other to enhance understanding through verbalizing and visualizing.
- ✓ Relationships among mathematics skills and concepts are emphasized.

Grades Three through Six

- ✓ Program promotes the understanding and enjoyment of mathematics, and provides a high degree of motivation
- ✓ Students must have experiences to apply what they learned in interesting ways, and many opportunities to explore and experiment with new Ideas.
- ✓ The program must break away from the traditional teaching of rote isolated computation skills, and instead must strive for the development of higher level thinking skills.
- ✓

Grades Six through Eight

- ✓ The middle and junior high school grades are important transitional years in the mathematics curriculum. Most students are developing the ability to think more abstractly. They have had many experiences applying skills and strategies for solving problems. The mathematics program should include:
- ✓ Strengthening of previously taught skills with a focus on greater depth of all strands.
- ✓ A variety of application assignments that incorporate more than one strand.
- ✓ More opportunities for thinking independently, solving complex problems, and working in small groups.
- ✓ Preparation for the more rigorous content and expectations of high school programs.

KNOWING MATHEMATICS IS DOING MATHEMATICS.

A person gathers, discovers, or creates knowledge in the course of some activity having a purpose. This active process is different from mastering of concepts and procedures. It is clear that the fundamental concepts and procedures from some branches of mathematics should be known by all students, established concepts and procedures can be relied on as fixed variables in a setting in which other variables are unknown, but instruction should persistently emphasize "doing" rather than "knowing that'.

Some aspect of doing mathematics have changed in the last decades!

Mathematics is a foundation for other disciplines and grows in direct proportion with its utility. The curriculum for all students must provide opportunities for them to develop an understanding of mathematical models, structures and simulations applicable to all disciplines.

Changes in technology and the broadening of the areas in which mathematics is applied have resulted in growth and change in the discipline of mathematics itself.

Technology

New technology has made calculations and graphing easier. It has changed the very nature of problems important to mathematics and methods' mathematicians use to investigate them. Because technology is changing mathematics and its use, it is believed that:

- ✓ calculators should be available to all students at all times – appropriate for their grade level
- ✓ a computer should be available in every classroom
- ✓ every student should have the opportunity to use a computer
- ✓ teachers should realize the computer as a tool for processing Information performing calculations to investigate and solve problems and can be as asset to students.
- ✓ students should learn how to use the internet (under guidance) in helping them solve problems or create new problems.
- ✓ the internet can help teachers add new interest to their teaching, and give them ideas for "making math fun" and exciting to learn

HOW SHOULD WE TEACH THE CURRICULUM?

The curriculum for each grade level must follow some guidelines. Although we may know the curriculum that is to be presented, we also need to know how to present the material to assure understanding.

Kindergarten -Grade 4

The K-4 Curriculum should:

- Be conceptually oriented – it should enable students to acquire clear concepts by being actively involved in their own learning.
- Enable students to construct, modify and integrate ideas
- Emphasize the developmental level of the students

- Allow students to think, reason and solve problems
- Emphasize the application of mathematics and help students see how it relates to everyday life.
- Include technology.

Grades 5-8

- Continue the sequential development of what was learned in K-4
- Motivate students to explore new ideas
- Include a broad range of topics which will help students communicate with and about mathematics.
- Technology should be an emphasis]
- Students should continue to be active learners with new and challenging situations introduced which require the use of their intellectual ability.

Students in grades 5-8 are unique. They are going through a time of change in intellectual, psychological, social and physical development. They are becoming more independent thinkers and their reasoning becomes more abstract. Concrete materials still must have a place in their development, as they need to be able to construct their own meanings, as they assimilate and accommodate new information. They are able to do more writing in mathematics, and can use language to help them clarify their thinking and reporting. They begin to question what they are learning and may require great depth in explanations of what they are being asked to learn.

All grade levels should include the students' cultural backgrounds in the learning experience. Black or Hispanic students, for example, may find the development of mathematical ideas in their culture of great interest. Teachers must be sensitive to the fact that students bring very different everyday experiences to the mathematics classroom. The way in which a student from an urban environment and a student from a suburban or rural environment interact with a problem may be very different.

Students will perform better and learn more in a caring environment in which they feel free to explore mathematical ideas, ask questions, discuss, and make their own mistakes. By listening to students' Ideas and encouraging them to listen to one another, one can establish an atmosphere in which there is mutual respect.

MATHEMATICS IS SEQUENTIAL

Activities that begin at the Kindergarten level build on experiences the students have had prior to entering school. The 1989 National Council of the Teachers of Mathematics Standards stated: "It takes careful planning to create a curriculum that capitalizes on children's intuitive insights and language in selecting and teaching mathematical skills and ideas. It is clear that children's intellectual, social, and emotional development should guide the kind of mathematical experiences they should have in light of the overall goals for learning mathematics. The notion of a developmentally appropriate curriculum is an important one."

A developmentally appropriate curriculum encourages the exploration of a wide variety of mathematical ideals in such a way that students retain their enjoyment of, and curiosity about mathematics. It incorporates real world contexts, children's experiences, and children's language in developing ideas. It recognizes that students need considerable time to construct sound understandings and develop the ability to reason and communicate mathematically. It looks beyond what students appear to know, to determine how they think about ideas. It provides repeated contact with important ideas in varying contexts throughout the year, and from year to year.

There are some skills that we consider pre-number skills. These skills are developed through varied hands-on experiences. Different manipulatives and materials that are readily available in the classroom, playground, home, etc., can be used in helping students develop these skills (multiple embodiments). However, In order to develop these skills, the student needs to be aware of the attributes of the manipulative.

Knowing number names and number words is not a prerequisite to beginning readiness activities. Students first need to be able to recognize the attributes of the objects, so they can manipulate the objects.

Communicating Mathematics

Children will not communicate mathematics until they have experienced mathematics. They must have a variety of experiences, and be given time to think about what they are doing. Once they have the appropriate experiences and feel they understand their actions, they will be able to communicate about mathematics through the medium of language. They must be able to share ideas and talk about issues.

Writing as a medium should not be used during the initial stages. Once the children feel comfortable with the models, and their actions on the models, they will begin to communicate and write. The type of communication will depend upon the teacher's instruction. Mathematical understanding grows as children explore ways to represent the ideas. Manipulatives remain important throughout the process. As children progress in their thinking, their means of communication will also progress. As time progresses, the children will reach the ultimate goal of recording in the abstract form.

Communication can be in the form of the spoken language, the written language, the graphic representation or the active mode. Each mode has two aspects - the receptive and the expressive. The mathematics classroom has a tendency to use the receptive mode, and the expressive mode is left to the paper and pencil. Students are required to work silently and individually. They feel that mathematics is nothing but symbols, terminology and manipulation.

Teacher language is another means of communication. The teacher needs to:

- Use the appropriate language in instruction to help children succeed;
- help children learn the relevance of mathematics;
- value the children's contribution;
- use appropriate vocabulary, syntax and symbols;
- use appropriate questioning, discussing and exploring Ideas.

Language and Mathematics

Language and communication are essential elements in all learning. Children learn by investigating the world around them. However, their active involvement is more than touch, feel, taste, observe, collect, pull apart and put together the things around them. It is thinking, reflecting, organizing and applying what has been learned to other situations. As children observe, the language elements are interwoven with the skills and processes of observing, comparing, sorting, classifying, making hypotheses, testing, designing, constructing, and evaluating.

The language of mathematics is a means of communicating mathematics. It involves more than giving a correct answer; it involves being able to tell why they have obtained an answer whether correct or incorrect. In our "new new mathematics" the children need to be able to construct their own meaning for mathematics, not depend on the print that has been provided for them. They will depend on the materials they use as their vehicle for thinking and communicating.

Whole language may be another way of helping children learn mathematics. Although whole language is being questioned as a means of teaching reading, it can still be used as another means of teaching mathematics. Whole language enables the student to write about what they are doing prior to assigning abstract numbers to the operation. Irons and Irons, in the 1989 NCTM yearbook, have stated that students need a broad range of language experiences in order to help them develop the fullest understanding of mathematical concepts. They also feel that children's knowledge and excitement about mathematics grow when situations are provided to encourage discussion about their learning. This allows students to expand their own strategies and build new ones.

Irons and Irons feel that learning experiences which encourage the development of language in mathematics courses can occur in four stages, with each stage having three phases.

<u>Stage 1: Child's Language</u>

The natural language a child uses to describe the concept in a familiar situation, often a real world story. The three phases within the stage include (1) modeling the situation, (2) creating new situations, and (3) sharing that situation (this is the same for all stages).

Stage 2: Material Language

The new language that might be used with concrete and pictorial materials as the child acts out or represents the real world story.

Stage 3: Mathematical Language

The use of a few words to record the language that describes the action of the materials. This stage leads to using more specific mathematical language.

Stage 4: Symbolic Language

The use of mathematical symbols as an even shorter way of recording action. (p. 86)

In the child's language stage, the students are writing stories that reflect their actions. This could be considered the creative writing stage of developing a mathematical concept. The material language is the stage during which they are writing to answer a particular question that is asked. This should be the stage where students begin doing addition and subtraction. They can write a story about grouping together or taking away from a set. At the mathematical language stage, they combine words and numbers, and in the symbolic language stage, everything is written in symbols. The fourth stage, symbolic language, is suggested by the Irons' as the first stage for most mathematical teaching.

Since more teachers favor teaching reading than mathematics, it may be one method of getting teachers more interested in teaching mathematics. Teachers, however, do need to think seriously about using language as a means of teaching mathematics. Maybe it is the key to success in teaching mathematical concepts.

Writing in the Mathematics Classroom

Writing is another part of the language curriculum that is finding its way into the mathematics classroom. It has been said that children learn to write by writing. The question is: Do children learn mathematics by writing? There is one danger in children learning mathematics through writing - children who write in the mathematics class, must have guidance regarding what they are writing. A child asked to write in his journal may say, "I had fun in mathematics today." This statement does not tell what has been learned in mathematics – it only expresses a feeling. We want children to record and communicate mathematical ideas and processes. A good assessment of a child's knowledge of mathematics is to have them write, in words, how they would solve a problem. They do not need to produce the answer, but tell how they would get the answer.

Writing helps children clarify and extend their understanding. They write about what they are doing and what they have learned. They must organize their material and reflect on their knowledge and understanding. Ask children to write about how they solved problems. If children's writing produces only peripheral writing, more directions need to be given to help the children focus.

Writing using mathematical ideas and concepts need not be factual. Stories provide for imaginative writing with strong mathematical content. In fact when students begin to work on word problems, they need to be the authors of the problems. This will help solve the mystery of word problems, and enable them to develop a positive attitude toward a part of math that is dreaded by most students. It will also encourage them to become creative problem solvers in different situations. As students write their word problems, they can become creative in their problems, and be able to share their expertise with the others in the classroom.

Reading and Mathematics

Reading plus mathematics is a two-way process: reading contributes to mathematics, and mathematics contributes to understanding what we read. Many of our ESL students find reading is one area that causes difficulty in learning new mathematics.

If we integrate mathematics and reading, we set up contexts for problem solving. Reading can provide the initial stimulus for an investigation, illustrate or broaden a unit of work, and synthesize the learning at the end of a unit.

Literature and Mathematics

Mathematics and literature help us understand our world. Literature books and poetry have given teachers a new view of mathematics. Through using literature books, a teacher can create an environment in the classroom where children can interact with the book. A wide variety of books are available for use in the mathematics classroom; not just those which are mathematical (Appendix A). In selecting children's books to use in the mathematics classroom, the teacher needs to ask, "What mathematics can be taught with this book?". Literature books are not limited to the primary grades; they are available for all grade levels.

Many literature books help the students become a part of the story. Once the teacher determines how the class can interact with the story, they develop lessons that integrate mathematics into other areas of the curriculum. Literature books can also be used as a means of developing manipulatives that will enable children to interact with the story, and investigate many mathematics concepts.

Literature books also enable a teacher to find a book that is of interest to the students. Students do not have to be able to read the book; it can be read by the teacher, and a mathematics lesson developed around the book.

The combination of mathematics and language, used in conjunction with opportunities for talk and discussion, allows children to grapple with mathematical concepts in a meaningful context. By placing tasks in a context that was familiar and that made sense to the children, researchers found young children were able to perform at levels that Piaget claimed they were unable to attain. Books extend and develop children's ideas of the world. The familiarity with the book or the story gives the child structure within which to explore mathematics. There are opportunities for problem-solving, patterns for ordering activities, and classification, as well as other mathematics skills. Literature can provide a link between the complexity of the world around us and the highly structured discipline of mathematics.

Mathematics and literature are both used to order the world around us

- both are concerned with classification
- both use problem solving
- both look at relationships
- both involves patterns

Mathematics should not be imposed upon a work of literature, as this would defeat the purpose of integrating the subject areas. But rather, the mathematics should flow from and be a natural part of the book. By using literature in this way, we can stress the activities that may follow. Reading or sharing of book activities must be a natural development of the book.

TEACHING FOR UNDERSTANDING

Students' understanding of the basic concepts is of primary importance in the mathematics curriculum. The understanding is more important than the students' ability to memorize a series of basic facts. Many students have come to view mathematics as a series of recipes to be memorized or computational procedures, with the goal being to calculate the correct answer. The assumption is that students who can perform an arithmetic computation understand the operation, and know when to apply it. Teaching for understanding emphasizes the relationship among mathematical concepts and skills, and leads students to approach mathematics with a common sense attitude, understanding not only how, but why, skills are applied.

Students who are taught to understand the structure and logic of mathematics have more flexibility and are able to recall, adapt, or even recreate rules, because they see the larger pattern.

Teaching for understanding does not mean that students should not learn mathematical rules and procedures. It does mean that students learn and practice these rules and procedures in contexts and are able to construct meaning from them.

Thinking

The teaching of mathematics has not changed much over the years. The methods used by your grandparents and great-grandparents are still being used today. However, the emphasis is different. We are in a technological age and must adapt instruction to fit the time and the place. Not many of us would want to go back to the horse and buggy days, but we are expected to teach mathematics as they did in the "good old days".

Reinforcement of Concepts and Skills

Many of the basic mathematical concepts are learned in primary grades. Basic concepts spiral from the early grades to the later grades. We begin teaching addition in Kindergarten and First Grade, and continue (in different forms) throughout high school and college. If a student does not have a firm grasp of the previous skills, they will not learn the new skills required, or be able to apply these skills to higher level thinking skills.

Students cannot be expected to learn new skills and concepts if they have not met the prerequisite skills. They need to see mathematics as a series of related concepts, not disjointed topics which they must memorize.

Homework assignments are often used to provide the students with more homework. We never know if the parents or the students complete the homework. Homework should be designed to involve the parents and the family members. Homework should not be repetitive drill on material already learned. Isolated practice is no more appropriate at school than at home.

USE OF MANIPULATIVES (CONCRETE MATERIALS)

Concrete materials or manipulatives provide a way for students to connect their understandings with the written algorithm. Manipulatives should be used to introduce new concepts, all the way through high school. Manipulatives are more than play things – they are the connection which helps students develop their own meaning and put their understanding into language. The teacher's role is to help the students make the transition from the concrete to the symbolic – bridge the gap.

Question: *Why do we need to use manipulatives to teach mathematics?*

- manipulatives are a model of the abstract algorithm
- manipulatives help students develop visual images which are crucial for later mathematical development
- manipulatives help children develop the basic concepts through hands-on experiences
- manipulatives help students internalize their own learning
- manipulatives take the fear out of wrong answers
- manipulatives build understanding and help make sense out of what the teacher is telling them to do.
- manipulatives allow for experimentation of basic concepts
- manipulatives allow the students to memorize basic facts without the drudgery of flash cards, etc.
- manipulatives build problem solving skills
- manipulative have no set language – all students regardless of their first language make sense out of manipulatives

Before students can internalize a concept, they must interact with the materials and discuss alternative approaches or explanations. Students also need to be able to talk about their interactions. This can be accomplished through cooperative groups.

MULTICULTURAL EDUCATION

We live in a society which has students from many countries. Although mathematics is a universal language, and most symbols are universal, the language involved in mathematics is not universal. We teach only the English version of mathematics in this country. Algorithms are not universal! Many times we find that students coming from other countries have been taught to do their algorithms differently, and have been successful in mathematics. However, once they come into our classrooms, they are forced to do them the way we are doing them. We ask students to show their work, without thinking that in some countries this is not allowed. The student must do mental math and keep the answers in mind while completing the problems.

Students who enter our classrooms may be forced to redo math they did two years prior in their own country. These students may become bored with the math class. They are no longer challenged in math. However, math is the subject that makes them feel comfortable, as they understand what to do.

Punishment is a major issue with students who have had their education in other countries. Students from some countries talk about the punishment they received when they either gave an incorrect answer or when they did not have their homework completed. Many state they have been hit when they did not perform in their classroom.

Another issue with students from the other countries is they are used to a highly structured classroom, where they only speak when spoken to. They are very much the passive observers in the classroom. When they enter our classrooms, they are asked to talk and discuss problems. They are encouraged to work in groups and 'talk' in the classroom. This is totally against what they have experienced.

Each culture has its own unique customs. In some countries it is disrespectful to look a person in the eye. In other countries the focus is on individual efforts, not group efforts, and group work is seen as "cheating"; in other countries, much of the work completed is group work. In some countries it is a sign of respect to call the teacher "teacher", while some teachers state they must be called by their name.

TEACHERS MAKE THE DIFFERENCE

What is the Teacher's Responsibility?

The teacher is the gatekeeper of the curriculum, the facilitator of learning.

The teacher must:
- be aware of cultural backgrounds
- allow time for ~"why", "try", and "I"
- stress oral language even in mathematics

- make sure practice does not precede discovery
- introduce vocabulary

The teacher

- uses the appropriate vocabulary
- is a role model
- develops the intuitive vocabulary of student
- doesn't talk down to students
- doesn't assume students hear what you say
  (You say: "The cat is in the tree."
  They hear: "Cat there")
- doesn't assume the student understands the language being used
- provides experiences that will help students understand their actions if learning is to be permanent,
- makes sure the student understands
- has formal and informal assessment, at each level.
- plans instruction for each child's strength, cognitive style, language background and preferred input mode.
- involves students in activities that reflect an understanding of their learning needs and abilities.

Assess Your Feelings Regarding Mathematics

The way you feel about mathematics may affect the way you teach mathematics. A common statement regarding mathematics is "I was never good at mathematics, and I avoided taking as many mathematics classes as I could." Many teachers have this same attitude, and have been spared the pain of teaching mathematics by hiding behind the textbook. They could give timed tests and assign pages from the textbook. If there were pages they did not understand, they would skip them. Concepts such as geometry were left out of the curriculum altogether.

Teachers do not have to be mathematicians In order to teach mathematics. They need to know the mathematics they will be teaching, and they need to know the methods of teaching mathematics. Introducing mathematics through manipulatives and activities will help teachers recreate their interest in mathematics. Mathematics teaching does not need to be dull and boring; it can be fun and exciting.

Textbook adoptions should focus on making mathematics exciting for all students. It should involve students as the major participants, and the teacher as the 'facilitator of learning': It is the teachers job to set the stage, where the students will be the major players – where they will be able to investigate, make predictions, and make mathematics relevant to their own interests."

How does the way I feel about mathematics affect the way I teach?

The teaching of mathematics is very important. It is more than opening a textbook and doing some algorithms. It is developing a program that meets the needs of the students in the classroom. It is teaching the class as individuals, not as a single average child who has no problems in mathematics. It will require the teacher to do preplanning and assess the students' mathematical ability. It will require the teacher to know his/her students, i.e., are there ESL students? What is their mathematics background? What was taught in the previous grade? What are their strengths and weaknesses; and how can we create an interest in what we are teaching?

The teaching of mathematics has been highly criticized the past few years. It is either the new textbook adoption, or the testing system used, or the lack of basic skills in the curriculum. Many of the people criticizing mathematics have the attitude, that the way I learned mathematics was good enough for me and I learned it, therefore, it is good enough for the students in today's classrooms." They do not realize that our society is changing, and we must prepare the students to live in today's society. They must understand what they are learning in order to be able to apply it to other situations. The lack of the ability to apply learning has caused application scores on tests to be lower than computation.

Our mathematics scores are compared to mathematics scores from other countries. However, they fail to realize that we teach all of our students, not just the select few. In many countries, children in the early grades are tracked as to the type of curriculum they will receive. If they are college bound, they take more advanced classes. If they are not college bound, they will be in other categories and their curriculum adjusted accordingly.

Flexibility of Instruction

Daily lessons should be structured to address the diverse needs of students through a program of ongoing diagnosis and assessment of each student. Flexible groupings, cooperative learning, whole group instruction, and individualized instruction, aid in the flexibility of instruction. Lesson plans should also be flexible, allowing for immediate adjustments to the daily plans.

Questioning and Responding

The way the teacher responds to students' answers can influence the answer. The effective teacher's responses are those that promote further thinking. Some examples are:

- Allow time after students' responses
- Accept students' responses
- Ask for clarification
- Suggest sources for further Information

What Experiences have Children had Prior to Entering School?

A crucial consideration here is many students are bilingual and have shared their experiences in their native tongue not English. These students also come from a variety of cultures and each culture has its own expectation of the student.

Many times, students enter our schools and are required to learn mathematics in English, which has little or no meaning to them. Even though they have had the same experiences, their understanding is in a different language. The language is the manifestation of culture, and creates past and present reality for the individual. Language is important in concept development, and a determinant of concept formation. Language is a set of symbols used to label concepts and provide a quick reference and meaning to concepts. Inconsistency in language might cause confusion in the student's mind, and delay his/her understanding. English Learners come from different cultures, and are products of the cultural system of their respective countries.

Most students have had the following experiences in their native country.

- shared activities with parents and siblings
- played games
- explored alone
- constantly interacted with their environment
- developed their own attitude In regard to math
- have become a complex individual
- have been exposed to different cultural background.

Think before you teach!

The teachers' attitude has an affect on the mathematical future of their students.

Chapter 2

Planning for Mathematics Instruction

MATHEMATICS THROUGHOUT THE GRADES

Planning for mathematics instruction is more than the words "open your books and turn to page x". Planning means being able to look at the page and decide what the student will need to know in order to be successful in completing the page. It means assessing the learning which has taken place. It means knowing the sequential development of a mathematical concept; and it means being able to motivate students to want to learn. It means asking the question:" *What does the student need to know in order to be successful on this page?*" Without analysis and planning there will be no teaching! Telling students to open their books and do a page is not planning or teaching. It is a way of getting around one's obligation to students, or a way to let the students know that the teacher does not know the content.

There are three types of lessons which evolve in the mathematics classroom.

- Introductory lesson – the first time a lesson has been taught
- Maintenance – a continuation or practice of a previous lesson
- Review lesson – a lesson that was taught previously, but the students need to revisit the lesson to retain the concept.

Before teachers can plan for lessons, they must know which type of lesson they will be teaching. If it is an introductory lesson, they will need to spend time with manipulatives in order to develop an understanding of the lesson. This may take more time than a maintenance or review lesson. The time spent at this level will assure success for the students. A review lesson should be ongoing throughout the year. Teachers can never teach a concept and forget it.

Once the type of lesson is determined, planning is next. One of the most important components of mathematics for any grade is the planning. Planning **WILL** make the difference between success and failure. Planning gives you the insight to determine what you want the students to do, and affords you the opportunity to reflect on the lesson. Despite the amount of time one spends planning, there are always lessons which are not successful. We need to reflect on our failures or less successful lessons and let them serve as a guide for future lessons. Remember: we learn from our failures. One of the most important things to do prior to teaching a lesson is to ask yourself the following questions:

- *What do the students need to know in order to learn this lesson?*
- *How will I know which students have the prerequisite skills?*
- *How will I motivate the students to participate – even those who are not interested?*

At the end of the lesson, you need to reflect on your lesson by answering the following questions:

- *Was it successful? Why?*
- *Would I change anything?*
- *What was the best thing about the lesson?*
- *Did the students participate?*

Teachers can not answer these questions if they do not plan carefully. Try answering these questions when your lesson is "open your book and do x pages".

During their first years of teaching teachers may have lessons that fail – lessons where they say to themselves "I can't teach mathematics", However, failure is part of the learning process. All teachers have been through this phase even veteran teachers. Failure for one lesson does not mean you are not a good teacher. Look at less successful lessons and ask yourself, "What should I change to make this lesson better?" Reflect on what you know; ask those around you; and use the techniques to either reteach the lesson, or review certain parts. If at any time you do not know the answer to a problem or situation, be truthful and let the students know. Tell them, "I will find out and let you know." Don't forget to let them know the next day.

NEW EMPHASIS IN MATHEMATICS

An attempt to change mathematics has been undertaken by the teachers of mathematics. Beginning in 1980's a group of teachers, professors, and researchers began to talk about changes, and gave suggestions as to what they thought would help students learn mathematics. Talk continued in the 1990's and is still continuing at the turn of the century. Teachers have joined the effort, and want to make mathematics relevant and empower each student in their ability to do mathematics. There are also the vocal few who want to keep this from happening.

The national and state mathematics leaders have written two documents that are important to the teaching of mathematics. The NCTM (National Council of Teachers of Mathematics) Principles and Standards for School Mathematics and the Mathematics Framework for California Public Schools are our guidelines. They give suggestions for content and instruction. The 2000 NCTM Principles and Standards state that the standards will play a guiding role in the improvement of mathematics education. They state a common foundation is needed for the mathematics that all students must learn. They also state students must be able to use math in their everyday life and everyday work. They feel we must focus on:

- Mathematics for life
- Mathematics as part of cultural heritage
- Mathematics for the workplace
- Mathematics for the scientific and technical community.

They have given a new emphasis to six principles for the continual improvement of mathematics education in the classroom, schools and educational systems.

Equity

- Equity requires high expectations and worthwhile opportunities for all
- Equity means accommodating differences to help everyone learn mathematics – if understanding is assessed only in English, some students' mathematical proficiency may not be assessed
- Equity requires resources and support for all classrooms and students

Curriculum

- A mathematics curriculum should be coherent
- A mathematics curriculum should focus on important mathematics
- A mathematics curriculum should be well articulated across the grades

Teaching

- Effective teaching requires knowing and understanding mathematics, students as learners, and pedagogical strategies
- Effective teaching required a challenging and supportive classroom learning environment
- Effective teaching requires continually seeking improvement

Learning

- Learning mathematics with understanding is essential
- Students can learn mathematics with understanding

Assessment

- Assessment should enhance students' learning
- Assessment is a valuable tool for making instructional decisions

Technology
- Technology enhances mathematics learning
- Technology supports effective mathematics teaching
- Technology influences what mathematics is taught.

As we move to include the new changes, we must not forget the previous changes that were suggested. In order for students to reach their potential, teachers must make certain they continue to:

- Use manipulative
- Provide for cooperative work
- Discuss mathematics
- Use questioning techniques
- Ask students to justify their thinking
- Use writing in mathematics
- Use a problem solving approach to mathematics
- Integrate content
- Use calculators and computers

As we move forward, we need to know the content is being taught to all students. Teachers will need to continue to teach for the needs of the next generation which means teachers will have to be knowledgeable about resources; they will need to continue to grow through professional development; their curriculum must be mathematically rich to assure students learn with understanding; introduce a variety of topics; use technology; and have a curriculum rich in problem solving. They need to be able to communicate mathematics, and in turn give their students the opportunity to communicate mathematics. In doing so they will enhance the mathematics learning of all the students in their classroom. Following these guidelines will give each student the opportunity to value mathematics and be engaged in actively learning it.

Teachers will also need to change their focus on what is important in mathematics. Mathematics has always focused on the number of right and wrong answers, with grades being assigned according to the number correct. Teachers have failed to consider the students who are the consumers of mathematics. We need to have a vision wherein all students have a solid mathematics curriculum and have knowledgeable teachers who can integrate assessment, education policies that enhance and support learning, classrooms with ready access to technology, and a commitment to both equity and excellence. Teachers will need to look at assessment and determine what the student does or does not know – not who has the right or wrong answers.

ASSESSMENT THROUGH DIAGNOSTIC TEACHING

Diagnostic teaching enables the teacher to look at the individual students and determine which skills and concepts they understand. It gives them a starting point for teaching, so students will feel successful in mathematics. If you use a textbook and ask every child in the class to do the same page, you will have those who speed through the page, those who struggle with it, but gradually complete the page, and those who do not understand anything on the page. Your classroom is not one student. Diagnostic teaching provides for all the students in your room.

What Is Diagnostic Teaching?

Diagnostic teaching is simply teaching to student's strengths, while improving their weaknesses. If students are having difficulty with double-digit operations, check to see if they understand single-digit operations, and then build on their strength with single-digit numbers.

One problem teachers have in this type of teaching is that they must make allowances for individual differences. The teachers must have an assessment tool with which they can determine students' strengths and weaknesses. This tool does not necessarily mean a standardized test - it can be looking at an assignment and assessing the errors on a page in a book, or a worksheet, or by observing the student while working. When the teacher does a careful assessment of the error patterns, s/he then decides where to begin teaching. If a student has no errors, the teacher will then give them activities that require critical thinking or problem-solving. Something that will not force them to repeat the tedious boring tasks they already have mastered.

Another point of diagnostic teaching is to give credit for those parts of the problem that are correct. Just because students have missed a digit does not mean they do not know how to do the problems. In the traditional classroom, if one digit is wrong, the entire problem is wrong, and no credit is given. This procedure sends the message that the student does not know the concept or skill.

If the teacher follows the "one problem one answer" frame of mind, students will score lower on each and every assignment plus it can cause frustration which could lead to anxiety. A question mark, or some other mark, after the wrong digit will tell the students they need to look at the error and assess their answer. This allows them to find their own errors and correct them. This procedure requires more skill than getting all of the answers correct. If they successfully find their mistake and correct it, they should be given full or partial credit for the problem. If the student is not able to find the error, then it is time for intervention.

Calculators can be an aid in helping students discover their own errors. After the student has completed his/her work, they check it with the calculator. If the calculator answer differs from their answer, they will need to redo the problem to discover the error. Many times this requires in-depth knowledge of the skill or concept.

Many times a mathematics lesson is taught to the total group without concern for individual needs. In reading we usually group children in order to enhance learning. Why can't we do this in mathematics?

PIAGET'S SUGGESTION FOR LEARNING A NEW CONCEPT

Piaget increased teachers' awareness of students and how students learn. He suggested students go through different stages in learning a concept, and are developmentally different at each stage. All students will pass through the stages, but not necessarily at the same age.

He felt students needed to be provided with many opportunities in order to interact with objects in their environment - either natural, or those placed there by the teachers. Through this interaction with objects, students experience new situations and are able to relate these new situations to those already in their hierarchy of ideas.

Piaget says we are constantly assimilating and accommodating new information. Just when we think we know something, we learn something new and have to fit it into our existing structures, a process he calls "disequilibrium".

> **Example**: When students learn their basic facts and can do simple operations, they are happy and feel they can do mathematics. Multidigit operations are then introduced, which cause disequilibrium. Disequilibrium helps students fatten their concept bank.

As we move from one concept to another, we must be careful to begin at the readiness level and build from there. If we do not, gaps in our concept bank could result, thus causing learning difficulties.

PIAGET suggests that students learn new concepts through:

- REPETITION

    Young students, in particular, enjoy doing the same thing over and over and over once they can do it. This repetition of an activity helps the student learn more about the concept.

- MONOLOGUE (talking aloud)

    Many times we request quiet in the classroom. However, Piaget states that students must talk in order to learn a new concept.

- DUAL OR COLLECTIVE MONOLOGUE

    Talking aloud to others whether or not the other is listening.

Question: Is our demand for quietness hindering learning?

The child's development level changes as the student mature. Through maturity they are able to:

- adapt information

- exchange thoughts
- comment on activities
- address remarks to others as part of interaction
- tell or ask others to do things
- address remarks to others as part of general knowledge
- question
- answer

Stages of Development

Piaget says students in the primary grades are at a "preoperational" stage. This stage usually lasts until age 7. However, age is not the determiner of the stages, but the developmental level.

Question: What are preoperational students like?

- naturally curious
- reason on the basis of how things look
- have their own personal point of view
- mind is like a still camera
- see things one at a time
- cannot move back and forth and keep the original attributes in mind
- Cat - Rat (see the word cat and the word rat, but cannot discriminate that the only difference is the beginning letter)
- 2 + 3 = 5 (see the 2 and the 3 as individual attributes. Once they are combined, they only see the 5; the 2 and 3 are no longer there)

    It is like taking three separate pictures

    Picture one   2 🚌🚌
    Picture two   3 🚌🚌🚌
    Picture three   5 🚌🚌🚌🚌🚌

- does not have attending or comparing capabilities
- language - talking to themselves is the key to development of thinking and language ability
- language develops at birth and continues throughout life
- language growth develops rapidly before age 8
- egocentric in thought and action
- thinks everything has a reason or purpose
- perceptually bound

- makes judgments on how things look
- if student is faced with objects with multiple attributes - long round yellow pencil - see what catches eye first.
- cannot conserve - decision is made on what they see
- no reversibility - cannot see original without original being there
- can't sympathize or empathize
- sharing Is difficult

What are Concrete Operational students like?

As students mature they enter the "Concrete Operational" stage. This stage begins at 7 years old and continues until 11 or 12.

- begins to think logically
- reverses thought processes
- can distinguish that the difference between cat and rat is the beginning letter or sound
- can put things together and take them apart – mind is like a video camera it can take a picture and rewind to show the reverse operation

    $2+3=5$; therefore, $5 – 2 = 3$ or $5 -3 = 2$
- order objects according to several attributes - i.e. height, weight, etc.
- begins to understand the operations
- can conserve quantity
- notices reversibility:

    notices $2 + 3$ and $3 + 2$ are reverse operations. When they begin to notice this, they are ready to begin memorizing the facts. Before this they would be required to memorize all 100 basic facts.

Students come to school with a wide variety of experiences, and many come with a partial understanding of many concepts, such as counting and adding, dividing and fractions.

SUGGESTED SEQUENCE FOR DEVELOPMENT OF ANY CONCEPT

Piaget states: Concept development begins with recognition of the characteristics of objects. As we introduce new concepts, there are certain steps that should be followed if we want the student to understand the concept, and gradually be able to use the concept for problem solving. Rote learning does not always insure a problem solving ability.

1. Free Play

According to early childhood specialists and many psychologists, a period of free play is crucial to the development of concepts at any age. Free play is a period where students are allowed to use materials in any way they want. During free play they begin to abstract ideas. Students discover relationships whether in the realm of numbers, language, or science.

> Example: Give the students some blocks and they will explore the many different ways to manipulate them. They will build towers, buildings, cities, etc. Through these actions, the students learn the relationship and characteristics of the blocks

Once they have made sense out of their manipulation, they are ready for instruction.

2. Instruction

While free play provides students with hands-on experiences, the teacher must help them make sense out of their free play. The first step of instruction is to introduce the vocabulary, just as in a reading class, and then build on the vocabulary before students are asked to do operations with the concept.

Example: Blocks

<u>Vocabulary</u> *add, take away, altogether, equal*

The student has had a period of free play and has an understanding of the characteristics of the blocks. The teacher is then able to direct the students to do activities that will help them understand the concept. A discussion will help the students abstract the ideas of the different blocks

2a. Bridging the gap between the concrete and abstract

Many times when we use manipulatives to teach a concept we go immediately to the written form of the problem. When we do this, the student usually does not see the connection between the manipulative and the abstract algorithm. A bridge needs to be made to let the student know the written algorithm is related to the manipulative.

Example: Blocks
- *Teacher asks student to place two blocks on the table*
- *Student places two blocks on table and records a 2 (on paper, magic slate, etc.)*
- *Teacher asks student to place three more blocks on table*
- *Student places three blocks on the table and records a 3*
- *Teacher asks student how many blocks they have altogether; the student combines the 3 and the 2 and records a 5*
- *The student can then move on to making their own problems. They are provided a "random number generator" (die) or given cards with the numbers 0-9 on them. They roll the die or draw a number and match blocks to the number. They do the same thing again to get the second number. They then place the two groups*

together to get the answer. This provides practice with creating their own individual algorithms.

This action helps the student understand the relationship between the model and the number. The student should have lots of practice recording their actions.

2b. Visualize the Process

Once students have used manipulatives, they will begin to develop mental pictures of their manipulatives. Teachers need to help them develop their visualization skills. Only after they have developed their visualization skills will they give up their manipulatives. Remember, the students give up the manipulatives when they are ready - when visual images have replaced the actual objects – the teacher does not take the manipulatives away.

3. Practice (at least 20 minutes per day on new concepts)

Practice is probably the most controversial stage in mathematics. In fact, this is the stage where most mathematics instruction begins. Once students have been introduced to the concept, know the vocabulary, and have had some directed practice with the concept, they are ready to practice the concept. Practice does not mean page after page of worksheets, or 50 pages out of the book. Practice means a limited amount of work spread out to give the student a sense of the meaning of the concept. Practice should also provide maintenance of previously taught concepts. Practice for each new concept should be approximately 15 minutes each day. This, plus a review of previously learned concepts, should make up the child's required assignment for mathematics.

4. Ultimate Goal – Problem Solving

All students will understand the skill or concept, and be able to do abstract work and the problems. There are many ways we can help the student maintain their ultimate goal.

Worksheets

There are many different types of worksheets available that add variety to practice. Some are coloring worksheets, some give a secret code, others reveal a pattern when the correct answers are recorded, and others have only computations.

Worksheets can also be designed that require the format used in the achievement tests; thus the testing format would not be foreign to the student.

Textbook Pages

Pages from the textbook provide a natural practice for the students. The biggest problem is teachers sometimes have a tendency to overuse the textbook. They feel the only way they know that students understand the concepts presented is to give them an outlandish amount of practice. More doesn't necessarily lead to greater understanding; many times more reinforces the wrong answers.

Games

Games provide drill and practice in a fun and relaxing atmosphere. Games can be teacher made, or purchased at one of the many teacher supply stores or ordered through a teacher supply catalog or be on the computer. Games can also be student- made. They love to make games for others to play. Games also force some students to learn their mathematics as they do not want their peers to know they have not mastered the concept.

EVERY STUDENT SHOULD HAVE AN OPPORTUNITY TO PLAY GAMES, NOT JUST THOSE WHO FINISH THEIR REQUIRED WORK.

Computer or Internet Programs

Computers and internet programs add variety to practice. There are many computer programs available that can be used to reinforce the concept you are teaching. This is also true about internet games and activities. Even if you only have one computer in the classroom, use it with the class - make a schedule that assures each student has a turn at the computer. Computers are good for cooperative learning activities. Some students will even be able to do simple programs that will help them with drill and practice.

Internet programs also reinforce concepts. The teacher needs to have a listing of the sites available for the students to use, and to monitor them carefully. If the students have a computer at home, the teacher can share the sites with the parents, thus allowing for practice at home. The book "Mathematics on the Internet" gives many internet sites for all mathematical concepts and for all levels of mathematics.

USE VARIETY OF PRACTICE: a combination of textbook pages, worksheets, problem solving activities and games helps students learn mathematics.

PROBLEM SOLVING ACTIVITIES

Problem-solving is said to be the ultimate goal of mathematics. If students can use the knowledge they have learned to solve problems, they understand the concept. There are many everyday problems that can be given to students to make them aware of the practical application of mathematics, as well as, provide practice in using the concept in other ways than algorithms in the book or on paper. Many of the textbooks have suggested problem-solving activities listed. These can be done whenever the teacher feels they are relevant.

Some students may have trouble with completing pages and pages of mathematics, yet are great problem solvers. This means they may have the skills, but find the paper and pencil drudgery. Since the ultimate goal is problem solving, the teacher needs to make a decision as to what is important for the student.

Questioning

Another important component of a lesson is the questioning techniques. Teachers have a tendency to ask lots of questions. However, most questions require correct responses. Teachers need to develop questioning techniques that will enable the students to reach their maximum potential. There are two types of questions – information-seeking questions, and questions that test. Gaudier, Cullinan and Strickland (1993) stated some general guidelines for asking questions:

- "Don't ask too many questions, and ask questions that are brief and clear. Spend more time allowing children to talk than you spend talking.

- Wait at least 15 seconds after you have asked a question before you ask another. If your questions require thoughtful answers, you must give the students time to think

- Listen to what the children are saying, and ask questions that encourage them to amplify what they are saying. Sometimes simply repeating their words with a raised intonation is enough.

- Ask questions that require children to infer, predict, hypothesize, and evaluate as well as questions that ask them to recall, define. and compare or contrast.

- Ask questions that cause disequilibrium, that force children to question their own ideas and assumptions that help them become aware of their logic.

- Open-ended questions such as inferring, predicting, and evaluating, allow for a variety of answers. By definition, an open-ended question has no more than one right answer. When asking these questions, accept a variety of answers, and encourage several children to respond.

- Give children the opportunity to ask their own questions of you and of each other. Before they came to school, children learned a great deal about their world by asking questions.

- Allow them to continue to learn in this manner."

Questioning techniques, careful planning, and diagnostic teaching will empower all of the students in mathematics. It will help them understand, and they can ask questions to clarify their misunderstandings. All of these provide a classroom atmosphere conducive to learning.

Question: ***How do you plan a lesson using manipulatives?***

The steps for planning a lesson using manipulatives should follow Piaget's suggestions.

Step 1: Place the materials in a learning center for approximately one week prior to using them for a teaching lesson. Let the students explore the materials and use them to build, make patterns, etc. The restrictions will depend on the students in the class - no throwing or hitting will be allowed. (Once the procedures have been established and students understand them, the problems become non-existent).

Step 2: Plan the lesson to use the materials.

Objective: what do you want the students to do with the manipulatives?

Motivation: Many times the manipulatives will be enough motivation to begin the lesson. However, you may have to enhance the manipulatives with motivation such as a literature book.

Readiness: Play time with the manipulatives is a motivation. However, you still must determine the prerequisite skills needed to complete the task.

Instruction: Language and vocabulary developments are important during this time. If you have a bilingual classroom, you will need to determine if you are going to do a transition to English, or teach in the native language. Manipulatives use the language assigned to them.

Note: The type of lesson you do will be determined by the materials you use. In no case, should you move directly to a paper and pencil lesson. Discuss the manipulatives, ask the students to do an activity with the materials, i.e. sort them according to some attribute. Do several different activities with the students prior to assigning them independent work.

Step 3: Independent work or group investigation. The students are asked to do some type of independent work. This is the first time that paper and pencil activities should appear. However, they still do not need to appear at this point.

Step 4: Review the independent work. Clarify the points that have been misunderstood.

Step 5: Assign further work that will enhance the previous learning. If you are going to a textbook, be sure to "bridge the gap" between the manipulatives and the textbook. Have the students explain what they are going to do, in order to insure that they understand their assignment. Ask several questions of different students.

Assessment: Do a walkabout assessment after you have assigned a lesson. This will give you a clue as to whether your instruction met their goals or if re-teaching will be needed. Observe the students working individually or in groups to see if they understand the concept. Plan other lessons according to your observations. A test is not needed to determine if the students understand what was presented. The teacher using the material that is available, or textbook pages, can develop future lessons that compliment the assignment. You need to pick and choose the pages the students need to complete. Do not use the entire book.

HOW DO YOU PLAN FOR A TEXTBOOK LESSON?

A textbook will probably be your security blanket for the first couple of years of teaching. There are ways to plan for a textbook lesson which will make it interesting and empower the students. The steps are:

PreTeaching –
- Look at the pages in the textbook you will be teaching.
- Read the Teachers' manual and look at what the students are being asked to do.
- Check to make sure you understand the pages, and can answer any question that may be asked (if students asks a question you cannot answer, be truthful and tell them you will have to check further and will give them an answer the next day. Be sure to follow through).
- Do all of the problems you will be assigning to the students.
- Check to see what type of a lesson you will be teaching:
    Introductory lesson (the first time it has been introduced in the textbook at that grade level)
    Maintenance lessons (a continuation of a previous lesson with more practice or practice with a different idea).
    Review (a few problems for a concept that has been previously introduced and assumed mastered by the students).

Question: What do the students need to know in order to learn the concepts or skills on this page? (Do they need to know how to add, subtract, fractions, etc.? More than likely, they will need to know several skills.)

Once you have determined what they will need to know, it is time to prepare your lesson. (Lesson plans are personal and may differ with each lesson.

However, the general format is always there, even when forced to follow a district or school lesson plan format.

The Plan for a Textbook Lesson

The lesson plan is the tool for guiding us in our teaching. The lesson plan can be very detailed or it can be a few words written in a box. When teachers first begin teaching, or when they are teaching a new concept, they need to do detailed (not scripted) lesson plans. The detailed plans may be tedious, but they may mean the difference between success and failure.

Objective: What do you want them to learn from this lesson? This does not necessarily have to be a written objective, but it should constantly be in your mind. It should not say with s % of accuracy.

Motivation: Many students are "turned off" when it comes to mathematics. They do it because they have to get a grade, and many times will say, "How will this help me in life?" Your next question is: "How do I interest the students in what I am teaching?" Many students have "attitudes" about mathematics, and you have to determine how to make the attitude positive. You may want to relate it to something of interest to your group i.e., if you have boys interested in baseball, use it for a motivation. How do they figure earned run averages etc.? Literature books are also excellent motivators.

Readiness is a must for each and every lesson. You have to prepare them for what is being taught. Readiness is the beginning of all mathematics instruction. It is the level of total development that enables a student to learn a behavior.

Readiness also is the level where manipulatives are introduced. You will have to select manipulatives that are appropriate for the lesson and grade level. If your school does not have manipulatives, use some home-made ones. Beans are the cheapest manipulative around, and can be used at all grade levels.

When using manipulatives of any kind. You must have them available for a period of free play.

Instruction. Language and vocabulary development becomes a crucial step during direct instruction especially for second language learners. In order for the student to make sense out of their free play, they must understand the vocabulary. This is also a time when they must 'bridge the gap" between the manipulative and the written algorithm. Also, be aware of students who have had their instruction in a different country. Ask them about how they learned a concept, and then be aware of this in your own direct instruction. What questions can you ask to enhance their learning?

Maintenance: This can be accomplished by the use of textbook pages, computers, ditto sheets, problem solving etc. It is a time for them to practice the skills that have been taught. This is a good time to do a combination of individual and group work. Many times, if a student does not understand the concept or skill, group work will help insure an understanding.

- **Assessment** This does not necessarily mean testing it could mean observing the students to see if they understand the concept, or checking their portfolio to see if the error is consistent, or asking them to explain their answer. This is the time you decide whether remediation is necessary, or if the student needs enrichment activities. Your textbook will probably give you suggestions for both of these.

- **Problem-Solving**. This is our ultimate goal: to have the students be able to use the concepts in the lesson to solve any problem presented.

The above mentioned sequence is a suggested sequence. It will take time to assimilate and accommodate the information, but in time it will become automatic. Preparing for a mathematics lesson is not easy. It is time-consuming and requires lots of planning. Another step that must be considered in the lesson is the language of the students. If the students' language is other than English, a bridge must be made to their native language.

BE YOUR OWN DECISION MAKER AND CHOOSE THE PAGES AND PROBLEMS THAT REINFORCE WHAT YOU ARE TEACHING. DOING MORE PAGES DOES NOT ASSURE THE STUDENT WILL LEARN THE CONCEPT.

Each step builds on a previous step.

Mathematics is sequential. Each step builds on a previous step. In the United States we spiral the curriculum. Concepts begin in the primary grades and build from that stage throughout the students' entire mathematics career. When planning a lesson, you must be aware of the sequential development of a concept, not the grade level. Sometimes it is necessary to begin at a much lower level to make sure students understand the concept being taught.

The California Framework has identified strands, which need to be met at each grade level. The sequential development begins in Kindergarten and spirals throughout the grades. The content covered for each strand depends on the developmental level of the student, and what has been taught prior. As we do the activities, we need to ask ourselves: "What mathematics is involved in this lesson?" We also need to ask ourselves how each of the strands can be applied to what we are teaching. The strands are:

> Number sense
> Algebra and Functions
> Measurement and Geometry
> Statistics, Data Analysis, and Probability

Problem Solving and Mathematical Reasoning

The first step in each of these strands is to begin with manipulatives. This is the stage for "hands on" materials. After students become acquainted with and understand the concept via manipulatives, they move to the pictorial stage. At this stage they will use pictures of the objects rather than the objects themselves. The final stage will be the abstract stage, the writing of numbers and letters

An example of the spiral curriculum with the use of a manipulative is with the use of Geoblocks.

GEOBLOCK ACTIVITIES

Geoblocks are wooden blocks, which come in many shapes and sizes. They can be used for many activities. At the onset, geoblocks were part of the Kindergarten program, and were used for activities such as the block center. Although they were considered noisy by some teachers, they are useful in teaching geometry concepts. They are a good example of how a concept can be developed through the use of manipulatives.

Activities can be used in grades K-6 to develop the concept of geometry. The activities which can be completed with the use of the geoblocks depend on the teachers' own creative thinking.

Geoblocks are a multicultural manipulative where the language can be changed to any language. A manipulative such as geoblocks can be used to teach several grade levels. Geoblock activities should begin in Kindergarten – at the beginning stage of developing a concept. This stage enables them to internalize the concept. They geoblocks can be used to make pictures (tracing the sides and making the sides into a picture of some sort) at the simplest stage to the more difficult concept of making a jacket for the geoblocks. As they work with the blocks, they begin to identify the vocabulary needed to discuss the geoblock – faces, vertices, edges, etc. Once they are successful with the prerequisite skills for area, they can move into teaching area, perimeter and surface area. Each step builds on a previous step. Through the use of the blocks, the students discover the formulas they need in order to solve geometric problems. Through the use of geoblocks, we can integrate all five standards into lessons for all students.

When we work with the geoblocks, we need to determine several factors, which may not be seen as mathematical. The **first factor** needed by the student is the vocabulary. A "word wall" can have mathematical terms added to it, or if you don't have a word wall, a chart can be displayed with the various vocabulary words in the language of choice – or dual language.

The **second factor** is the lesson planning for the block activities. We need to consider several points when we plan lessons. The steps which must be identified in a lesson are:

    I.   Vocabulary – What words do the students need to know? What steps will you take to make sure the ESL students understand the vocabulary?

II. Motivation – How will you get the students interested in the activity?
III. Objective – What do you want the students to do?
IV. Skills needed to learn this lesson - What do the students need to know in order to complete this lesson?
V. Materials – What materials are needed to complete this lesson? (Organize the materials prior to the start of the lesson)
VI. Procedures – What do you want the students to do during the lesson? Will writing be required in this lesson? What study skills are used during this lesson? What will you do with the student whose language is other than English?
VII. Practice – What practice activities will you give the students to help assure understanding of the concept?
VIII. Assessment – How will you assess the activity?
IX. Closure – How will you bring closure to your lesson?

The **third factor** to consider in using the blocks is the organization of the classroom. In this particular activity, it will be best to break the class up into small groups and have each group complete the task. The tasks can be written on individual cards, or on paper. One means of setting up the activities is to have the students rotate from task to task. The tasks should be in different parts of the room to avoid confusion. The students need to understand the sequence of the activities, and what they should do if they complete a task prior to being ready for the next task. The other procedure would be to have all students do the activities at their own center. You would need to have enough **materials** so each group would have all materials. They would then be assigned a set of activities to complete, and given the sequence to complete the tasks.

There should be some type of recording going on with each activity. If you have very young students, you might want to use a walkabout assessment to determine if they understand the concept. As you are doing the assessment, ask questions which will help you make an assessment.

One of the most important factors to consider is "How much can my students complete in one class period?" Be sure you have enough to keep them busy, but not so much as to frustrate them. Lessons can be spread out over several days – remember, we need repetition to learn concepts. See Appendix B for a sequential lesson plan for Geoblocks.

Summary

Planning, questioning. and diagnostic teaching can help you become a successful teacher in your classroom. If students are failing, don't blame them - maybe it is you that is failing. When students fail, reflect back on what has been done. Did you plan for the lessons? Did you present the lesson in a manner in which all of the students could understand? Did you revisit the lesson and teach it in another manner? Remember, your students can only be as successful as you let them.

PLAN ALL OF YOUR LESSONS CAREFULLY
REFLECT ON YOUR LESSONS
YOUR STUDENTS DEPEND ON YOU!!!

Chapter 3

What is Assessment?

Assessment gives us a look at the students' knowledge base.

Most of us think of assessment as a test that we have to study for in order to obtain a grade or an achievement test that tells us how we rank against certain populations or tests that tells us if we are capable of doing a certain job such as the California Basic Education Skills Test (CBEST). We are part of a testing society. Our future depends on how we take tests and the scores we receive. Many individuals have had their dreams shattered just because they could not pass a test. Many courses rely on a test to determine the grade we receive in the class regardless of the work completed prior to the exam.

Standardized tests begin in the primary grades. Many district give state tests, plus district test, plus national exams to tell how the student is doing in class. A one time calculation was completed to determine the amount of time spent testing at a particular grade level – the analysis was one month of each school year was spent testing. All of this could have been eliminated if they would have trusted the teacher's judgment regarding the students' strengths and weaknesses. Not only is crucial teaching time reduced by testing, but the cost takes away from other materials that could be used in teaching.

Assessment is changing. We are beginning to think that although students can pass tests, many cannot apply the knowledge they have obtained. They are the products of the rote learning society. They have not had to think, only reply. A change in emphasis is difficult for some teachers. Teachers will have to be facilitators of learning rather than providers of knowledge thus causing them to make changes in their teaching styles. They will have to review their philosophy of teaching and be willing to give up their old habits of teach and test.

Teachers need to follow some of the suggestions of the NCTM. Although this emphasizes math, it can be extended across the curriculum.

The NCTM Principles and Standards states that assessment is an integral part of mathematics instruction and contributes significantly to the student's mathematics learning. It is stated that "Assessment should be more than merely a test at the end of instruction to see how students perform under special conditions; it should be an integral part of instruction that informs and guides teachers as they make instructional decisions." (P.22) Assessment should be routine in the classroom – not an interruption in teaching. The point is also made that formal assessment provides only one point of view. A variety of assessments should be ongoing in the classroom – interactive journals, portfolios, student work, observations, and answering questions. If the teacher collects evidence from more than one point of view, s/he will have a more accurate picture of the mathematical ability of the student.

If a teacher wants to find out information regarding students' strengths and weaknesses, they will find several methods of collecting data for an ongoing assessment. When a student is seen struggling with a concept or skill, the teacher needs to talk to the student and ask them to explain what they are doing. Many times the student does not realize they are having difficulty until it is brought to their attention. Once the teacher has collected the information, then a conference can be held to remediate the problem.

The teacher now needs to ask, "What will I do with this information?" Most tests that are given are scored and filed away. District tests, which are suppose to give information regarding a student, are filed with little attention given to the results. Good teachers do not need tests to tell them students' strengths and weaknesses.

There has been a history of myths regarding mathematics and other subject areas. The NCTM Assessment Standards for Mathematics list several myths regarding mathematics (p.5-6).

1. Learning mathematics means mastering a fixed set of basic skills; therefore mathematics tests should focus on whether students have mastered these basic skills.
2. Problems and application come only after mastery of skills.
3. First we teach, and then we test.
4. Student learns only by imitation and memorization.
5. There is almost always a single right answer to a mathematics problem.
6. Objective, multiple-choice tests are the only valid and reliable indicators of quality mathematical performance.
7. The purpose of assessment is to determine which students "have it" and which do not, and then to assign grades and placements accordingly.
8. Objective multiple-choice tests are the best way to measure the most important ideas in mathematics.
9. Class grade assignments should be based on the bell shaped curve.
10. In the classroom, only the teacher can adequately evaluate a student's progress.
11. Alternative forms of assessment are less objective than traditional forms of testing and simple the latest fad in ducking education accountability.

If teachers were asked to agree or disagree with these myths, a majority will probably strongly agree. These myths have been the background of mathematics teaching and testing for our parents and grandparents, and yes even us.

We need to take a look at assessment and ask ourselves what we really want to get from assessment. If we believe in the myths, we may be missing a great deal of information regarding our students.

Assessment and instruction must be closely linked. Assessment may be the teacher's test to determine if they have presented the information in a manner in which the students will be able to apply it. In this manner, a good teacher is constantly assessing the students work. They are "in tune" with what the students are doing, and the errors they are experiencing. Assessment is ongoing in the classroom, and instruction is not stopped to give a test. Assessment means that students are free to try different methods of solving problems and are not afraid they will receive a lower mark. Assessment is giving students credit for what they do know and building on that knowledge. While assessment is interested in the mathematical skills the student has, it is more interested in the application of those skills - a timed test will not tell if a student can apply the basic facts. Assessment opens up communication between the teacher, the student and among the students. Assessment aids learning and measures their mathematical knowledge and power.

The National Council of Teachers of Mathematics and the State of California have suggested some alternatives to the traditional assessment.

1. Open ended tasks
2. Student Observation
3. Portfolios.

Whatever method used should take into consideration the characteristics of the student. This is particularly important for early years when students are just beginning to develop their mathematical power. Their assessment should be closely tied to physical models. Assessment tasks that allow them to use materials are better indicators of learning. At this stage, we can ask the students to "tell me about the problem" or "explain to me how you got that answer". If they are having difficulty with a concept or skill ask them to show you how they obtained the answer. Another method found successful with young children was placing a question mark beside answers, which were not correct. Students were told if they could prove they were correct, they would be given credit. This eliminated the fear of failure. If a method such as this is used, the parents need to be informed since all papers will be perfect. Pencils can be used for correcting papers, as the mark could then be erased there was no fear of those ugly red marks on the paper. With older students, given them credit for the part of the problem they did correctly don't check only right or wrong.

Writing across the curriculum is now an accepted method of teaching. Writing can be one of the best assessment tools we have. A teacher can read a student's verification of an answer and immediately know if they have a misconception of the Idea.

Alternative Assessment

Student Mathematical Products

Student products or works that students have generated may include writing in the form of journals or open-ended questions, videotapes, audiotapes, computer demonstrations, dramatic performances, bulletin boards, debates, student conferences, presentations, student designs and inventions, investigation reports, mathematical models the list is almost endless" (P. 6 Assessment Alternatives in Mathematics)

Student Portfolios

Teachers and their students should be allowed to choose most of the items to include in their portfolios, since it gives a good indication of what is valued. Occasionally it may be desirable, for the sake of comparisons, for some outside agency to ask for inclusion of a certain type of item, but this should be the exception. If possible, teachers and students should be able to present and explain their own portfolios to outside observers.

A portfolio might include:

- written descriptions of the results of practical or mathematical investigations
- pictures and dictated reports from younger children
- extended analysis of problem situations and investigations v1descriptions and diagrams of problem solving processes
- statistical studies and graphics representation
- Reports of investigations of major mathematical ideas such as relationships between function, coordinate graphs, arithmetic, algebra and geometry
- Responses to open
- ended questions or homework problems, group reports and photographs of student projects
- copies of awards or prizes
- video, audio, and computer
- generated examples of student work
- 'other material based on project ideas developed with colleagues

(P. 8 Assessment Alternatives in Mathematics)

Writing in Mathematics

Communication has become important as we move into the thinking era. Writing requires understanding. Writing requires the students to make connections between what they have learned and what they are learning. Writing must be woven into mathematics.

Investigations in Mathematics

"One of the best ways to assure the connection between instruction and assessment is to embed assessment into instruction. When students become involved in practical or mathematical investigations, assessment can become natural and invisible. During the investigation, assessment activities or questions can be presented to students without their being aware of any difference between assessment and other classroom work. Investigations may be related to other curricular areas or just mathematics

Investigations can be:

- identify and define a problem and what they already know
- collect needed information
- organize the information and look for patterns
- Miscues, review, revise and explain results
- persist, looking for more information if needed
- produce a quality product or report."

(p. 13 Assessment Alternatives)

Open-ended Questions

An open-ended question is one in which the student Is given a situation and is asked to communicate most cases, to write a response it may range from simply asking a student to show the work connected with an addition problem to complex situations that require formulating hypotheses, explaining mathematical situations, writing directions creating related problems, making generalizations and, so on. Questions may be more or less open depending on how many restricting or directions are included."

The advantages of open-ended questions:

- recognize the essential points of the problem involved organize and interpret information
- report results in words, diagrams, charts, or graphs
- use appropriate mathematical language and representations
- write for a given audience
- make generalizations
- understand basic concepts
- clarify and express their own thinking.

Scoring open-ended questions is usually holistic:

- the papers are sorted into stacks, the stacks divided into two levels each. A "rubric" or a description of the requirements for varying degrees of success in responding to open ended questions maybe pre-defined or may be created as a result of reviewing a number of papers. If pre-defined, it should allow for the unusual responses that are often seen in open investigative work by students." (P. 19 Assessment Alternatives)

An ongoing assessment is the use of observation and interviews. Teachers who are aware of their students are constantly observing their actions within the classroom. A natural extension of the observation is an interview, or talking to the student. Although we could have a plan in mind, many times an informal observation or interview will supply us with a wealth of information. Good questioning techniques will add to our Information bank regarding the students.

Grading

With the use of alternative methods of assessment, grades become a problem, especially if a letter grade is required.

The Portfolios give us an indication of the students' growth. Through observation, we can determine if the student has met the requirements for a specific grade. The open-ended questions are usually scored through a "rubric", therefore giving us more concrete information with which to assign a grade. If alternative assessment is used, the teacher will have to find a way to assign a grade based on the student's effort in the classroom, and the improvement they have made during the year.

Alternative Assessment will not be easy. It is suggested that a teacher start slowly, and do alternative assessment for one subject, or one part of the subject, until they are comfortable with the process.

Writing across the curriculum and portfolios are excellent ways to begin alternative assessment. The portfolio gives us a wealth of information and allows us to keep the student's work for the year, and possibly pass it on to next year's teacher. This will eliminate the new teacher's question as to what the student learned the previous year.

Alternative Assessment may be our answer to helping all students learn mathematics, and may help us eliminate the criticism that our students can not do mathematics.

Chapter 4
MULTICULTURAL EDUCATION

Mathematics is multicultural! The mathematics in today's classroom and textbooks are a result of what was discovered centuries ago in other countries. This applies not only to the math taught in the United States, but also to the math taught throughout the world. The numbers may be different and the language may be different but the end results remain the same. Test the students in your class and determine which math processes they were taught.

Chinese Culture

As we look at the cultures that were the most influential in mathematics, the Chinese would probably be first - with the majority of their influence being felt in higher-level mathematics. They had a place value system and used zero as a placeholder. Their system was a multiplicative base ten system, which employed nine numbers, and additional symbols for place value components of ten. Some of their figures date to the third century BC. The Chinese calculated by means of sticks laid on a table – the beginning of the abacus.

The Chinese were also influential with their astronomical system. In 2852-2738 BC they changed the zodiac into 28 animals (which we still have today). This led to the necessity for knowledge of measuring time and angles. Another important contribution of the Chinese was their books. One of their greatest books was The Nine Sections. This book was held in high esteem for many centuries. Topics found in this book indicated they were the pioneers in the early science of mathematics.

The Chinese were also known for their activities, such as the "Magic Squares" – an activity that is popular in present-day classrooms. Pascal's triangle is similar to the magic squares, and has found its way back into the classrooms.

Hindu-Arabic Culture

The Hindu – Arabic system gave us the current number system that we use. They adapted this system from Europe, even though it came from an arithmetic book written in India. It was first translated into Arabic, and then Latin. This translation preceded the printing of books in Europe. The system was named for the Hindus, who may have invented it, and for the Arabs who transmitted it to Europe. The early system had no zero and was a non-positional. The zero and positional notation may have been introduced in India sometime around 800 AD.

The Hindu-Arabic system assigned values to whole numbers less than 20. In 1585 they extended it to the decimal system, when Simon Steven of Burgess found a scheme for making a poet separating the integers and the fractions.

In 825 AD the Hindu system became a positional system, and had a zero value. In 850 AD, Hindu mathematicians wrote the following statement:

> *A number multiplied by zero is zero and that number remains unchanged which is divided by, added to or diminished by zero.*

This contains the core concept of zero as the identity element for addition and subtraction; however, it contains a gross error namely, that division by zero has the same effect in addition and subtraction. Three hundred years ago it was stated that division by zero was "naught". From this time it was recognized that zero was a placeholder.

Influences of Other Cultures

Early tribes of Queensland - Base 2 systems – used as binary codes in today's computers

South African Tribes - Base 5 systems – popular, since we have 5 fingers

Babylonians - Base 60 system – used a positional principal

    They also had a subtractive system

    Early systems of weights and measures used Base 60

Egyptians - Earliest known numerals were inscribed in 3400 BC. Vertical strokes represented the numbers from one to nine. Individual symbols were used for successive powers of ten through millions.

Romans - Roman numbers were commonly used in bookkeeping. In the 1300's the use of Hindu-Arabic numbers was forbidden in European Cities because it was felt that "numbers" were easier to forget than the Roman numeral.

Mayan - One of their major contributions is the Base 20 system with positional notation and a special symbol for zero. They also gave us some areas of astronomy, the calendar, architecture and commerce.

Question: How does culture affect the teaching of Mathematics?

Today, we live in a multicultural society where many of our students have had experiences in schools in other countries. We know there is literature available which tells about schooling in other countries. However, in order to get a feeling for how math was taught in other countries, and what individuals remember of their mathematics training, preservice teachers were given the assignment of interviewing someone who had their schooling in another country. The following excerpts are from the interviews of these students. The age range of the students was from upper elementary aged through adults. Most of the interviews were from the Greater Los Angeles area where the minority populations were the majority. There were approximately 225 individuals from various countries interviewed. The students were given a series of questions to ask in their interviews. A summary of the questions is:

Question 1: Were textbooks used in your country?

Approximately 98% stated they used textbooks –with some variance between public and private schools. If textbooks were not used, the teacher would lecture at the blackboard and the students would copy what the teacher had written in their booklets. This would then become their homework. Many stated the textbooks were closely followed.

Question 2: Was homework part of the everyday curriculum?

An overwhelming 100% said "yes". The statements indicated that homework was expected, and was part of life. Some stated there was so much homework they did not have time left over for anything else. They also stated that their parents demanded homework. An interesting factor was the use of home work as a threat. All groups indicated there would be severe punishment for not completing homework – from swats with rulers to a mother coming to school to spank the child. Students also felt if they did not do their homework, it would reflect on their parents. Others indicated they felt homework was their job. Homework varied from 20 problems to 3 hours.

Question 3: Was it difficult to adjust to mathematics teaching in American schools?

Only two interviewed stated they were behind when they arrived in the United States – one was schooled in Bangladesh and the other was from a remote area of Mexico. The majority stated that they were one to two years ahead of their American counterparts. They felt that math was the thing that helped them keep their self-esteem and succeed in American schools. Many also stated they repeated materials they had learned two years prior in their own country. Others indicated they lost some of their math skills since they were no longer challenged to use higher-level mathematics, and were bored with the math they were being taught.

One student who went from the United States to Japan stated she was way behind when she went to Japan and had to work hard to catch up to the students there. When she returned to the United States, she was way ahead of the other students.

Question 4: Did your parents help with your homework?

There was more variance with this question than with all of the rest. The indication was parents did not actually help with the homework, but they were aware of the homework and made sure it was completed. If there was a problem, tutors were hired to help the students. Asian and Middle Eastern students talked about having tutors for mathematics, whether or not the parents were able to help. The Latinos, especially those from Mexico, stated that it was the teacher's job to help and make sure the students understood the homework.

Many stated their parents did not have the educational background to help with homework. Many of the parents had left school in the elementary grades. Others stated that although they had tutors in their homeland, the expense of moving to the United States was so great there was no money to hire tutors. These same students said they could not ask for parental help with homework in their new country, as their parents did not know the new language. They had to depend on a peer or helper or just figure it out on their own.

Question 5: Are there any math words that would cause children from your country to have problems?

Most stated language in general proved to be the problem. However, several did suggest that teachers should use only one term when introducing a word or concept, i.e., use either "add" or "plus", not both. The same is true for all operations. Another suggestion was to make a chart listing the words in English and the words in the child's first language.

It was also interesting that many students who studied English as a second language stated they had a difficult time understanding the American accent. Many knew "book English".

One Armenian interviewee suggested that words such as trading could be confusing since the translation may mean something totally different.

In Holland the word for equal is "is", so instead of saying 2 + 2 equals 4, they will read it as 2 + 2 is four.

Question 6: What could we, as teachers, do to help students from your country adjust to the mathematics teaching in America?

The main concern was that the teacher knows how math was taught in that student's country. Everyone had to be prepared to answer either orally or on a test. Failure to do so met with corporal punishment. Parents expected students to be prepared, and in order to avoid bringing disgrace to the parent; students completed their homework. Homework was student work, and they felt in order to learn they must have a lot of homework.

Many parents hired tutors to work with the students, especially if the student did not understand what the teacher was presenting. They were obedient students who did as they were told. Their main objective was to follow the rules. Cooperative learning, asking questions, and making choices were not included in their daily education. Fear was always there. Many feared not being able to pass the required test that would determine their future. They also wanted to pass the test so as not to bring disrespect to their parents, who valued education and expected them to become successful citizens.

The students felt due to the "lax and laid back" atmosphere in many American classrooms, students disrespected their teacher; the ability to ask questions and to question the teacher; freedom of choice; the lack of homework (which had kept them busy many hours in their own country); extracurricular activities; and repeating materials they had already learned made the adjustments to math in the United States even more difficult. They stated that time helps the adjustment, but teachers need to be part of the adjustment to ensure that these students will continue to learn and be challenged in their new environment.

Those interviewed from the Middle East (Armenia, Iran, Egypt, Lebanon & Israel) stated there could be severe consequences for not completing homework. They stated there was strict discipline in the schools, and the students may learn in fear. Praise for work was not part of the school day. More emphasis was placed on the boys' learning rather than the girls. Math was a complex, rigid, and competitive course, where memorization was the rule. Some stated school met 6 days a week, with homework every day - including the 7th day! Classes consisted of recitation and memorization. They also had a different way of doing division.

The students who went to school in European countries stated math was tougher and highly competitive. They had lots of memorization and recitation and higher-level math skills were taught. Several stated they used manipulatives to learn. It was also stated that in Germany students had to pass tests to determine which schools they would attend. German students had four years of elementary school, and then they went to one of three schools depending on how they did in the elementary school. Those who were college-bound went to "Real Schule" for 6 years, and after that "gymnasium" for 3 years or to a specialized synasium where they would only study one subject. The other track was the Haupt schule or "dummy school" where students learned the skills to become an apprentice and they were looked down upon. Their school year was from August 1st to July 15th, Monday through Friday, and every other Saturday.

South African students used British textbooks, and most of the work was from the book. Memorization was stressed. They stated that the American accent caused a language problem.

The Latino countries stated that classes were more advanced. They had tons of homework and they had to know their math or else. Memorization was required and the teacher was looked on as a second parent. Students called their teacher "teacher" as a sign of respect. They reported corporal punishment was used in the schools, and if they failed three subjects they would have to repeat the grade. They could repeat the grades as many as two or three times. They did a lot of choral repetition of the facts and the teacher was the only one who talked in the classroom. They said the children do not look people in the eyes and they may appear to be timid and shy. The students coming to school in the United Stated may be confused when they see other students not following the teachers' directions. Education is a "big deal" in these countries, as many do not attend school beyond the elementary years. The students also reported that the first two years of math in this country was mostly review.

Students who had had their schooling in the Asian countries stated they were overwhelmed with the amount of information they were forced to learn. They stated the teacher corrected worksheets and handed them back with no explanation. They were timed-tested constantly. There was no talking in the classroom.

Students who had some of their school in Korea stated math in the United States is less demanding and homework was given on Saturday. They also stated that in Korea, they were only required to listen, while in the United States they are required to listen and speak. They also felt math was much more enjoyable in the United States. They also were required to work alone; therefore, group learning may be difficult for them in the United States. They also reported that teachers should be aware that corporal punishment was used when students did not complete their homework, failed to behave or achieve adequate performance. They felt students studied for fear of negative consequences. During vacation time they worked in daily journals and prepared materials for the next semester. They also stated that Math was the one subject that helped them gain their self-esteem back in this country. They gained confidence in their ability to succeed. They also stated that fractions were different in Korea.

2/3

American way: "two thirds"

Korean Way: "three two" or "three under two"

The denominator is read first.

Students from Vietnam stated that memorization was the rule. They held lots of competitions in the class. They stated that division was taught differently, and that multiplication facts must be known by third grade. They also stated that "pd" was the abbreviation for pounds, not LBS. The method of punishment for failure to complete homework, as well as, classroom discipline, was to kneel down and the ruler was used to hit the hands. Students were expected to learn at least one other language. They also stated that the teacher's explanation for what needed to be done was better in Vietnam than in the United States.

The Japanese parents were very concerned with their progress. They put pressure on the students to perform, and many had private tutors, as the parents couldn't keep up with the academic materials.

In Hong Kong the students had to master each level and were not just passed from level to level. The subject is taught so as to be acquired quickly, without much explanation. Age or grade did not make a difference. Math was taught in Chinese until 6th grade and English in Jr. and Sr. High. Those who graduated from private schools with high scores on examinations were more likely to reach the top of the socio-economic ladder. The climb up the social ladder began in elementary school. Examinations were the way of life, and the higher the scores on the examinations, the higher the place in society. Parents were so involved in education they hired tutors to help students pass.

Students who have had their schooling in El Salvador may have lived in violence. Education was important in the country, and parents were highly motivated. They expected excellence, achievement and respect.

Those interviewed gave suggestions for Americans who wish to help students adjust to their American Schools:

- Be aware of the differences in algorithms and accept their ways of doing them
- Don't stereotype students, i.e. all Asians are good in mathematics
- Encourage students to be more creative and relaxed in class. Encourage them to speak freely.
- Explain the purpose of games.
- Use manipulative to help students understand the math concept and help them transfer their own concepts to American thinking
- Use the metric system
- Help children by talking to them individually, and find out about their background.
- Be sensitive to children with language barriers.
- Introduce one work at a time to students.
- Change learning behavior slowly – tell reasons.
- Introduce terminology – possibly a chart listing the terms in both languages.
- Keep eyes and ears open to different needs of the individual students from other countries.
- Help students build self-esteem and self-confidence. Students have low self esteem due to lack of ability to speak English.
- Some students have only been listeners – now we are asking them to be both listeners and speakers. Students have been influenced by the punishment system, and may have no reason to study in the absence of negative consequences.
- The adjustment to teaching methods are difficult and very different.
- Teachers should offer academic challenges to these students.
- Students have had to rely on mental mathematics and do not show their work

The students interviewed felt there were many advantages to the American Schools:

- Teachers are more in tune with how schools affect a person and their life They care more about feelings than performance,
- In America you learn what you choose – teachers are more helpful and approachable.
- American schools take their time teaching certain math information because children are not developmentally ready to learn it, whereas,

other students have been overwhelmed with information that they had to learn no matter what happened.
- Teachers in America do not "push the students as they did in other countries.
- The U.S. textbooks are basically self-explanatory.

Mathematics is multicultural. We need to be aware of how it was developed. We also need to take advantage of having students from different countries in our classroom and have them share their mathematical experiences with the rest of the class.

Teachers also need to be aware that many students have different methods of completing algorithms, and they need to be given credit for their answers whether it is the same method used by the teacher or not.

Be aware of the students' background and their math experiences in their own country.

Vocabulary development must be a part of each lesson for EL students.

Chapter 5

CONTROVERSY IN MATH EDUCATION

The teaching of mathematics has not been without controversy. As we look into the history of mathematics, we find we have cycled from understanding to rote memorization and back again. This controversy continues today. However, those who research the way children learn still maintain that understanding is more important than rote memorization. The goal is to empower all students in the area of mathematics, and in order to do this the teacher must be sure the students in the classroom understand what they are doing. The teacher is the keeper of the key, and works with the students daily. The teacher must have a good understanding of mathematics, and be able to present it in such a way as all students will be able to achieve.

The Principles and Standards for School Mathematics (NCTM) and the California Framework for Public Schools determine what is being taught in California. Although the two parallel each other, there are differences.

NCTM wants the mathematics classroom to be viewed as a place where students are thinking about and doing mathematics. They state students need to value mathematics, be confident in their own ability, becoming mathematical problem solvers, learn to communicate mathematically, and learn to reason mathematically. The sequential nature of mathematics enables students to progress through each stage to reach their potential. If students are to reach their potential, then teachers must prescribe what is happening in each classroom and to each student within the class.

NCTM's intention is to help teachers and policymakers expand their vision of effective mathematics teaching and learning to include a strong component about student's understanding of mathematics. They want the highest quality of mathematics instruction programs for all students. They want to set high goals for each of the students and find ways to help them attain these goals.

The NCTM feels there are three domains that play a role in the formation of educational standards:

- Research – helps determine what is possible for the students to learn and the sequence, which makes the learning more efficient.

- Professional Teaching Practice – the experience and observations of the classroom teachers, teacher educators, educational researchers, and mathematicians play a key role.

- Judgment of Professionals – research along with views of professionals in mathematics and mathematics education about societal needs for mathematical literacy, past practice, and expectations of the general public for school mathematics.

The NCTM Standards were written with all of the above taken into consideration. Once they drafted what they felt was an acceptable format, they made it available to approximately 20,000 members for their feedback. They listened to what those in the field had to say about the proposed Standards and made the necessary adjustments.

NCTM feels that the mathematical experiences the students have in school will enable them to become mathematically literate. How much they learn, and how they learn it, depend on the quantity and the quality of mathematical instruction. The materials used by the teacher will have a great impact on how much the students learn. Students who have a variety of opportunities to study "well-taught" mathematics are more likely to gain proficiency. This also means the teacher has high expectations for all students, regardless of ethnicity, socio-economic level, or prior schooling.

The NCTM has ten standards they feel are needed for students to become proficient in mathematics:

1. Number and Operations
2. Patterns, Functions and Algebra
3. Geometry and Spatial Sense
4. Measurement
5. Data Analysis, Statistics, and Probability
6. Problem Solving
7. Reasoning and Proof
8. Communication
9. Connections
10. Representation

A major emphasis in the teaching of mathematics must be: *"mathematics for all"*. The student's future may depend on the mathematics s/he gets in their schooling. Mathematics cannot be for the talented few. We must look at the issue of equity. Gender differences must be eliminated, and girls given the same opportunity as boys; the poor the same opportunity as the rich; and bilingual the same opportunity as monolingual. We must have mathematics to fit each student in the classroom, not just the select few - as many other countries do.

The Mathematics Framework Committee worked on the initial draft. The document states: "the draft prepared by this committee received a majority support, but not the unanimous support, of the committee members." The different math organizations throughout the state, and especially the California Mathematics Council, questioned some of the material in the document. It was stated that the Framework was sent out to 3000 people throughout the state.

The California Framework addresses two primary audiences: (1) educators; and (2) developers of instructional resources. The educators were those who were responsible for the day-to-day learning in the schools and responsible for curriculum and instruction. They also addressed audiences such as parents, community members (including business and civic leaders who have a vital stake in the success of California's students in mathematics). The following themes permeate the Framework:

- aligns the mathematics standards with curriculum and instruction
- a balanced mathematics curriculum
- addresses the needs of all learners
- the importance of mathematical reasoning
- stresses the importance of frequently assessing student progress toward achieving the standards
- avoids oversimplified guidance on either content or pedagogy in favor of guidelines on effective instruction derived from reliable research.

The goal was for the students to have a balanced mathematical program. They had defined the balance as having basic computational and procedural skills; conceptual understanding; and problem solving ability. In order for the students to reach this goal, there must be assessment, quality instruction, time allotted to do mathematics each day, instructional resources, classroom management and professional development. Also, administrators and the community must be involved.

The state has identified five strands that develop sequentially from Kindergarten through High School graduation. These strands are:

1. Algebra and functions;
2. Number Sense
3. Measurement and Geometry,
4. Statistics, data analysis and probability; and
5. Mathematical reasoning.

Suggested guidelines are given for each grade level, and in time the testing will reflect the guidelines.

Assessment is a major factor in teaching mathematics. The organizations which established the standards, state that all students will master or exceed world-class standards. To achieve this they must be continually challenged and given the opportunity to master increasingly complex and higher-level mathematical skills. There are three types of assessment: entry-level assessment; progress monitoring, and summative evaluation. These three assessments will provide a road map that leads to the mastery of all skills.

The latest documents from the NCTM emphasized the difference in each student's perception of mathematics and their thinking styles. This should be taken into consideration when teaching mathematics and assessing mathematics. One mode of response does not give an accurate indication of students' individual capabilities. For example timed tests reward the test takers and students who have a speedy response while it may have the reverse action on those students who are test anxious. Therefore, a variety of assessment activities must be used at all grade levels.

California's Framework has given a different answer for the timed tests. It states that timed tests play an essential role in measuring understanding – especially for the basic topics, where automaticity is required. They feel those who cannot answer quickly have a superficial understanding of the topic and they have not internalized the concept.

The document also states there must be mastery of the materials at each level. The only reliable method to determine this mastery remains the **timed test**.

Both organizations agree that we need proficiency in math. Both organizations want to see the mathematical abilities of their students rise, and they want to empower the students to excel in mathematics. However, they seem to disagree on how to accomplish their goals. One wants to set aside research and the opinions of those who are in the field daily, to focus on the "drill and kill" routine. This group assumes that more is better, and we must teach to the tests. One group has a large database for feedback, while the other group gains feedback from the select few. One group values the opinions of the educators, psychologists, etc., while the other seems to focus on what the business community prefers. Both groups have good ideas, and for students to succeed in mathematics, the teacher will have to be the decision-maker. Just because a student has had a "hands-on" approach to learning mathematics does not mean they cannot pass a timed test, or that they have superficial learning of a concept. In a few years, we will see the pendulum swing in a different direction, and we don't want this to hinder the learning of the students. We must teach so they understand, and we must assess daily the work of our students.

The goal for all teachers is: Help all students learn mathematics in a manner which will allow them to be in control of their mathematical learning.

REMEMBER: The teacher is the decision-maker, and is the person who knows the abilities of the students in the classroom and the content to be covered. The means of teaching these will depend upon the materials available for the teacher to use.

Chapter 6
Patterns, Function and Logic

Patterns, functions and logic can all be considered prerequisite skills to Algebra. In mathematics, many different topics fit into strands. Some topics will fit into one strand, while other topics fit into many strands. This happens to be the case for Patterns, Function and Logic. The California Framework has incorporated patterns and functions into a strand called Algebra and Functions and have added logic to the Logic and Language strand. NCTM Principles and Standards have placed patterns, relations and functions in the Algebra Strand.

Algebra has its historical roots in solving equations and has been seen as a Junior High or High School subject. Algebra should begin as early as first grade and build from the early experiences. Patterns, functions and logic are beginning prerequisite skills needed to help students understand algebra. Young children have experiences with classifying, and ordering objects, and making patterns. As children become more supplicated with patterns, they can predict a subsequent sequence for the pattern. Children also use an understanding of the properties of number to represent equations, i.e. through various hands-on experiences they view the equal sign as a means of balancing an equation or they use a number balance to see what is meant to equalize the equation. The variable can be introduced through what used to be called the "missing addend". Students who have experienced the concept of a balanced equation automatically can transfer their knowledge to adding a variable to the equation.

The Framework has stated that math has many unifying ideas that go across more than one strand, one of which is patterns (a prerequisite skill to algebra). Patterns play an important role in the elementary grades up through the middles school. To see how these are related, we need to know the definition of each one.

Patterns are regularities in events, shapes, designs and sets of numbers. These regularities enable us to sort and classify items as we recognize their attributes (characteristics). Identification of patterns helps students become aware of other structures. Patterns are everywhere in our environment. According to the Framework, patterns play a prominent role in all strands of mathematics, and its role is too important and too broad to link with one particular strand.

The Framework states: "Functions represent a way of generalizing a number pattern. Functions explore broadly the many kinds of relationships between quantities and the manner in which those relationships can be made explicit but not necessarily symbolic. (p.81) Functions are the backbone of mathematics but need not be formal, abstract, or imposing. Thoughtful experiences with a variety of functions, formal and informal, in early grades will help all students make functions part of their intellectual repertoire when the functions are met later on.

Family Math (p. 247) states "functions are a special kind of mathematical rule that gives a unique answer for each number." Functions can also be an input-output situation in the real world.

Logic looks at the connections or relationship between objects. Family Math states (p. 55) "Logical thinking is making good sense out of something, usually in an organized way. It includes sorting things by some characteristic or thinking ahead about what the results of an action might be.' Logical thinking requires experimentation. Students clarify and strengthen their reasoning by talking through their strategies.

Patterns help us become aware of the relationships between the objects. Through functions, we can find different relationships between the variables. Our answers will be unique to the function we are performing. Logic helps us make sense out of what we are doing. It enables us to organize information and sort out information and determine what the answer might be to our problem.

We need to help young students become aware of patterns and functions and engage them in logical thinking. Primary aged students can be introduced to these by providing them with activities that are appropriate for their developmental level.

Students need to have lots of experiences in classification, comparison, and seriation activities. If students do not have these skills, they will experience difficulty in mathematics. Each time a new concept is introduced, students need to have experiences with classification, comparison and seriation.

Classification requires the students to group objects according to an attribute. By changing the identifying attribute a set may be regrouped. Once the awareness has been developed, Students are ready to make patterns with the set either by copying or extending patterns. Through these activities, students develop the ability to think analytically. Verbalization helps them understand their actions and further reinforces the skills.

Concrete models are a must! Students cannot learn to sort and classify from pages in a textbook. The actual objects must be in front of them. Once they have completed the classification activities, they are ready to compare the likenesses and differences of various materials. At first, the concern is with the general characteristics – not the number. After students become aware of the general characteristics, they can sort according to number, without using number, i.e. are there more red or more yellow?

Seriation follows the classification and comparison. Some objects cannot be seriated (ordered) according to size. They may need to be placed in groups according to a different characteristic.

All students need many hands on experiences with various manipulatives. They need to able to verbalize what they are doing, and work in cooperative groups to reinforce the skills they know.

Patterns are everywhere. Patterns, as all other mathematics is sequential. In the primary grades students are encouraged to look for patterns, complete patterns, classify and organize patterns. These experiences enable the students to begin to see the relationship of patterns to their everyday life. As they continue to work with patterns, they will begin to express their actions in relation to mathematics. The skills developed at the primary level are prerequisite to the abstract thinking which is required in higher level skills of patterning in numbers, algebra, geometry and measurement.

From the earliest grades, the curriculum should give students opportunities to focus on regularities in events, shapes, designs, and sets of numbers. As they begin to recognize the regularities, they should be encouraged to verbalize what they have found and to expand on it through generalized patterns.

Pattern recognition involves concepts such as color, shape identification, direction, orientation, size and number relationships. All of these are required for extending patterns. Identifying the "cores" of patterns helps students become aware of the structures.

Students in the middle grades expand on what was introduced at the primary level. Patterns and functions is a central theme at the middle level. Students continue to recognize, describe and generalize patterns and begin to build patterns to predict real world behavior. They also begin to recognize the occurrences of regular and chaotic patterns which make the student more important. This helps the students develop the mathematical power and should instill an appreciation of the beauty of mathematics.

Patterns begin at the informal level and no matter what is introduced; it should be at the informal level. From this level it can expanded to analysis, representations, and generalization of functional relationships.

Students should be encouraged to observe and describe all sorts of patterns in the world around them; plowed fields, haystacks, architecture, paintings, leave on trees, spirals on pineapples, etc. As students mature, instruction efforts can move toward building a firm grasp of the interplay through tables of data, graphs, and algebraic expressions as ways of describing functions and solving problems.

The 1989 NCTM Standards suggest increased emphasis should be placed on pattern recognition and description use of variables to express relationships identifying and using functional relationships, developing and using tables, graphs and rules to describe situations, interpreting among different mathematical representations.

Classification Activities

One of the beginning activities for young students is classification - putting objects together using some type of an attribute. Classification, or sorting, activities can begin at school or at home. In the classroom, the teacher can begin to classify the students according to an attribute, i.e. the teacher places students wearing blue jeans in one line and those wearing white shoes in another line. The students then attempt to discover the differences.

What are attributes?

Attributes are characteristics that allow students to group objects. Through recognition of attributes the student can (1) classify, (2) match, (3) compare and (4) order objects. Students having difficulty recognizing attributes may have difficulty in recognizing the difference between the words cat and rat or the numerals 2 and 3.

Students come to school aware of certain attributes, but it is the teacher's responsibility to expand this awareness. This is also a time to help the bilingual student bridge from their native language to the English language.

ATTRIBUTES

Developing the awareness of attributes:

- Involve the students in various activities that will help them understand that an attribute is a characteristic.
- Use the students in the classroom and help them discover their own personal attributes. The first activities will deal with characteristics, which will allow the students to be sorted into one group or another; i.e. the type of shoes the student is wearing that day. The teacher might want to sort the students according to a certain attribute and help the students discover the attribute. Some of the attributes that can be used are:

➢ color of the hair
➢ type of shoes
➢ sex
➢ color of clothing
➢ color of socks
➢ color of shoes
➢ eyes

Activity 1: Personal Attributes

Vocabulary: *attribute*

Materials needed: *Students in your class*

Directions: *The teacher decides on an attribute (without announcing it to the students) and asks students to go to one row or another. The teacher might want to group them according to hair color, color of clothing, regular shoes or tennis shoes, color of shoes, socks or no socks, etc.*

Example: *Ask the students with blond hair to stand on the left side of the classroom and the students with brown hair to stand on the right side of the classroom.*

Question: *How are the students on the left side alike?*
   Are they different from those on the right side?
   How are they different?
   How are the students on the right side alike?

Note: *The emphasis on likenesses is extremely important due to the constant emphasis on differences. This helps students see we are alike than we are different.*

Once students have been grouped according to an attribute, list a word on the whiteboard (this could be in the form of a Venn diagram), and regroup the students according to another attribute. When the students have been regrouped several times, then refer to the words you have listed and introduce the word attribute.

Activity 2: Today's Special Student

(upper grade classes would have a more sophisticated title)

Vocabulary: *words to describe the special student of the day.*

Directions: *Write a sentence on the overhead or whiteboard to describe a student. The sentences placed on the board or overhead will help the class discover the special student for the day.*

- ✓ *The student is wearing blue jeans.*
- ✓ *The student has on white shoes.*
- ✓ *The student has blond hair.*
- ✓ *The student is wearing a t-shirt.*
- ✓ *The shirt is red.*

From these clues, the students should be able to determine the special student. If not add another clue. I.e. the student is in the 3rd row or in group 3.

Follow up activity: Use a cooperative learning group to have students identify the attribute of something in the classroom, i.e., their desk.

Activity 3: Who has my attribute?

Materials needed: *Ropes in the shape of a circle, labels, and students in the class.*

Directions: *The teacher makes the rope in the shape of a Venn diagram and places labels in the circle. A select group of students places themselves inside of the rope circles. Those, which have both attributes, must go into the intersection (this reinforces class inclusion or the fact you can have more than one attribute at the same time.)*

Suggested labels: *types of food liked, sports played, Words (hat, bat, cat. but, bit)*

Venn Diagrams

Once the students recognize they have one attribute or the other, then a series where they might have both attributes can be added. One way to do this is to use the Venn diagram. Ropes made in the shape of a circle can be used (they should be large enough to hold several students). Provide labels for the different circles such as "I have a dog and I have a cat" Those students that have a dog will get in one circle and those with a cat in another circle. What happens to those who have both a cat and a dog? A label, which says, "I have a cat and a dog." is added. At this time the students will have to exit the circles and the circles are moved together to allow for the intersection of the set. The labels are then realigned and the students get into the correct circle.

Activity 4: Find my classmate *(this activity would before 2nd graders on up)*

Vocabulary*: characteristics of students in the class, students' name*

Materials needed*: students in your classroom, cards describing the characteristics of the student in the classroom.*

Directions*: Divide the class into two groups. Have one group sit in a chair (or stand in front of the room, etc.) The other group will have a short description of one of the students sitting in the chair. They must match the characteristics with the student, and then stand behind that student. Once everyone has found his or her partner, each description is read, (For older students you could have each of them write a short description about themselves to use with this game.)*

ATTRIBUTES OF COMMERCIAL MATERIALS

After the students have had experience with personal attributes, they can then begin to use materials that are available in the classroom. They need to be able to observe any object and describe the characteristics of that object. Once they do this they will begin abstracting the object by describing the similarities and differences of the objects. Their personal experience provides them with a reference point from which to abstract other attributes.

Activity 1: What are the attributes of the classroom?

Materials needed*: cards for writing attributes, tape, objects in the room*

Directions*: Give the students some blank cards. Students move around the room identifying the attributes of objects within the classroom i.e. table. They write an attribute they observe and attach the card to the object.*

Extension*: Once the students have labeled the objects in the classroom, similarities and differences of classroom objects can be discussed. A list of similarities and differences can help them classify the objects.*

Upper grade students are given cards with many attributes on them matching the objects in the room or around the school. Upper grade students need to expand their horizons beyond the classroom. This could also be used as a homework assignment to look at the attributes of their homes.

> ### *Activity 2: Recognizing the characteristics of different attributes*
> **Materials needed**: *Unifix cubes, cuisenaire rods, attribute blocks, numeration beads, or any other material available.*
>
> **Directions**: *Spend a few minutes playing with the manipulative of your choice. What are the attributes of your manipulatives?*

> ### *Activity 3: Classification*
> **Materials needed**: *Same as in activity 1*
>
> **Directions**: *Classify your manipulative according to one characteristic (i.e. color). Classify your manipulative in a different manner.*

> ### *Activity 3: Comparing*
> **Materials needed**: *Different manipulatives*
>
> **Directions**: *How are the manipulatives alike and how are they different?*

After students are familiar with the attributes of various objects, and have a variety of experiences working with attributes, personalized attribute models can be introduced to the students. These attribute models can be related to reading, math, language, etc., or they can be something that is of special interest to your class. You can also make attributes that relate to the various holidays; hearts for Valentines Day, Christmas trees for Christmas, firecrackers for the fourth of July, etc. You decide on the pattern and the attributes and then make them out of material that is available.

LOGIC AND PROBLEM SOLVING WITH APPLES

Apples can help you jump into many logic and problem solving activities in your classroom without the drudgery of paper and pencil activities. Your students will have fun learning new and exciting mathematical concepts.

Although apples are used for this activity, individual attribute models can be made from many different material, i.e. cut paper, nuts and bolts, sticks, origami, etc.

The Apples were created using the theme from the book "The Giving Tree" by Shel Silverstein. Other apple books apples can also be used as a motivation or anticipatory set.

The apples have four general characteristics:

Size: Large or small; **Color**: Red or green; **Leaves**: Leaves or No leaves; **Worm**: Worm or no worm

These four characteristics determine the number of pieces in the attribute model

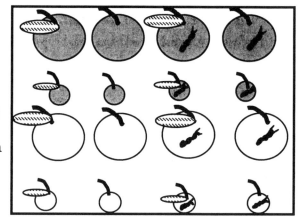

2 (sizes – large or small)
X **2** (colors–red (gray) or green (white)
X **2** (leaves or no leaves)
X **2** (worms or no worms), gives us a 16-piece model (2 x 2 x 2 X 2).

The attribute tree (See appendix C) shows the planning process for the apples. When you design your own model the attribute tree will help you plan the pieces to make. Place the name of your model on the single line on the left-hand side. This time it will be "apples". As you progress across the tree, you will add new characteristics to the previous characteristics. When you have completed the tree, you will only make the pieces in the last column (16 pieces).

These personally designed attribute models have been used with all aged students and with students in a multicultural setting. Many of them have been used with the student's native language and the translation made into the English language.

The activities and games mentioned in this chapter could be adapted to use with any model.

There are many activities to use with your apples. The activities can be expanded to the teaching of the basic facts. The activities enable the teacher to meet the individual needs of the class.

Vocabulary development is a must!

All activities must begin with vocabulary development. Vocabulary development is as important in mathematics as it is in reading; however, it is a step we usually fail to take. If you have bilingual students, you might want to introduce the vocabulary in both languages and help them bridge the gap between their native tongue and English.

The activities should always begin with a period of free play. Early childhood specialists state that free play is an important prerequisite to any activity. A period of free play allows the students to explore the models and begin to abstract the ideas of the models in their own way. Once they have had a period of free play, they are ready for directed activities. It is assumed that prior to using the Apples, that students have experienced the above mention attribute activities.

All of the following activities will use either a small set of the apples or an enlarged set which are easily visible from all around the classroom.

Activity 1: Identify the attributes of a given model
Vocabulary: attributes, alike, different, classify, sort, group

Materials needed: Attribute models similar to the Apple model.

Directions: Give the students a set of attribute pieces and let them explore the models for a few minutes. After they have explored the models, ask them to put the models together in some way they may match by size, or color, or shape, or any characteristics of their choice. There are many different ways to group the models, but the grouping is their choice. Ask the students how they grouped their models and list their groupings on the board.

Directed Instruction: Give directions to classify the models according to size, shape, design, color, etc. After each grouping, review and discuss the groupings.

Questions: How are the groups alike? How are the groups different? Can you sort or classify a different way? What do all of them have in common?

Activity 2: Describe your Apple
Vocabulary: Same as in activity 1

Materials needed: Either the individual Apple models or a large model of the Apples.

Directions: Pass out one Apple to each of 16 students. Ask each student to describe his or her Apple. As students describe their Apples, the remainder of the class listens. If a student with an Apple forgets an attribute, a student who has been a listener can take the Apple and add the missing attribute that student now joins those with the Apple.

Activity 3: Match the Apple
Materials: One set of Apples

Directions: 16 students are each given an Apple. The students are told to find any other Apple that is similar to theirs. They can find a match that is exactly the same except for a different size, etc. Once they have matched with another Apple, they must tell how they match.

Activity 4: Apple Train
Vocabulary: one different

Materials needed: Attribute model (or large models with strings so students can hang them around their neck); chairs used as imaginary tree placed in a line

Directions: Use the 16 Apples from Activity 2. Choose one child to sit on an Imaginary tree in the front of the classroom. Explain each tree in the train *can only have one attribute different. The students can add to either side of tree train (or according to the rules set by the teacher and students). Have students identify the Apple who could join the train next. As they join the train they must tell how they are one different. If a student has a one different attribute but does not come up, a listener may request the model of the Apple and join the train after telling the one different attribute.*

Learning Center Activity: After the students have played with the models, they can be placed in a learning center and used during free time.

This activity has been used successfully with 4 year olds.

Activity 5: I'm Thinking of an Apple

Materials needed: One set of Apples, Overhead projector

Directions: The teacher chooses an Apple and writes several attributes about that Apple (which the students do not see). The teacher asks the students to choose one of the Apples. The teacher then uncovers the first description of the Apple.

It is a large Apple

If the one the students have chosen does not match this, they then choose a different Apple.

The teacher continues until the class has discovered which Apple the teacher is describing. The language can be changed depending on the age and logic of the students involved.

Activity 6: Find the Apple

Vocabulary: *Words from the attribute model such as large, small, leaves (or no leaves) worms (or no worms)*

Materials needed: *One set of Apples and one set of cards describing the Apples. The first time the students do this, the language should be easy for them to understand:*

Example: *I am a small Apple. I am green. I have a leaf. I do not have a worm.*

The language of the cards can gradually be changed to increase thinking skill

.I am not a large Apple

My leaf is missing.

I have a worm.

My color is not green. Who am I? *For young students, you might want to use pictures of the objects such as leaves, worms, color or. large, or small.*

Directions: *Divide the class into two groups. Give half of the class an Apple and the other half a card describing an Apple. Have them find their partners.*

Learning Center: *Place the Apple and cards in a learning center for the students to use during free time.*

Variation: *The students observe the 15 Apples, and record which Apple they think is missing. They write their guess on a piece of paper and place it face down on the table. Once the missing Apple has been identified, a check is made to see who guessed the missing Apple*

Activity 7: The Case of the Missing Apple

Materials needed: *a set of Apples*

Directions: *Pass 15 of the 16 Apples out to the students. The students try to find out which Apple is missing.*

Activity 8: Estimation Which Apple Do I Have?

Materials needed: *Large envelope and a 16 piece Apple Attribute Model*

Directions: *Go around the room and ask the students if they can describe an Apple that is hiding in the envelope. If the student can describe the Apple, they get the Apple and go to the front of the classroom (this enables other students to see which Apples have been chosen and which ones are left).*

Activity 9: Who Am I?

Materials needed: *Large Apple Model (it should be easily seen from anywhere in the classroom), Set of cards listing the attributes of each Apple.*

Directions: *The large Apple models are placed on the backs of 16 students. These students must discover which Apple they have on their back. They can only ask questions which can be answered with a "yes" or "no". Once they know which Apple they have, they stand in the front of the room- they stand with their back to the class and tell which apple they have. The rest of the class gives some indication of whether they are right or wrong.*

Extension: *Write the description of each Apple on a card and once students have discovered which Apple is on their back, they pick up the card that describes their Apple.*

Activity 10: I'm Going to Drop from the Tree and I'm Going to Join My Friends...

Materials: *Apples, Imaginary Tree*

Directions: *Pass the Apples out to 16 students. Begin the story by telling about all of the apples that live on this very special Apple Tree. Give the description of the first "apple" student, who falls to the ground to join the apple game.*

The student (apple) who joined the apple game gives the description of the next "apple" to join the game. This continues until all of the apples have joined the Apple Game.

Activity 11 Apple Riddles

Materials needed: Apples, paper and pencil, bulletin board

Directions: Place the 16 Apples at the bottom of a bulletin board (or in a place where students can reach them). Each student writes a story about one of the Apples. The story is then displayed on a bulletin board and other students match the Apples to the story. The storywriter must check the Apple on the tree next to their story. If the Apple placed does not match the Apple being described, the storyteller replaces the Apple at the bottom of the bulletin board. Once all 16 have been identified, the stories are read and discussed.

ADDITIONAL APPLE ACTIVITIES

Activity 1 Apple Concentration

Materials needed: One set of pictures of apple attributes on playing cards or index cards and one set of cards describing each Apple.

Directions: Place all cards down (or a certain number of cards) and students take turns turning over the cards until they have a match. The winner is the student with the most matches.

Activity 2: Game Board Find the Apple

Materials needed: Game board, one set of Apples

Directions: The game board has shapes on it, and lines leading from the shapes (see board at end of handout). If there is only one line, the student will look for an Apple that has only one attribute difference from the first one. If two lines are coming from the previous shape, they must look for an attribute that has two different attributes.

Alternative version: Use cards, which have either one line on or two lines on them. The student chooses an Apple, and then draws a card. If the card has one different line on it, the student chooses an Apple with one different attribute. If the card has two lines on it, the student chooses an Apple with two different attributes different. Play continues until all Apples are used.

The cards can be varied, for students who need a challenge, by using three and four lines.

Activity 3: The Apple Picnic

Materials needed: Game Board (Large folder with a picnic scene): Use three dice. Students roll the three dice (two with attribute words, and one with the words "and", "or", "not") and make a logic sentence. They place an Apple on the picnic blanket which is described by the sentence.

Example: if the words "red", "worm", "or" are rolled, the player may place a red apple, or an apple with a worm, or a red apple with a worm on the picnic blanket.

Directions: Each student gets half of the Apples. Students roll the dice and match the dice to one Apple picture on the board. The first student to place all of their Apples on the picnic game board is the winner.

Variation: Randomly place an Apple on the picnic folder. Roll a die with only the numbers 1 and 2. If an '1' comes up, the student will place an Apple that has only one different attribute. If a 2 comes up, the student will place an Apple with two different attributes. (Cards with one line or 2 lines drawn on them can be used in place of the die.)

Activity 4: How Am I Different?

Materials needed: Apple cards

Directions: Place a set of Apple cards face down in the center of play. Each student takes turns turning over the two cards. They must state the differences between the two cards. If they state all of the differences, they keep both cards. If they cannot state all of the differences, the other student gets a turn. If all of the differences are not stated, the cards are placed back in the pile, and students take another turn.

Activity 5: Loop the Apples

Materials needed: Apples, ropes, cards with attribute words (large, small, red, green leaf, worm).

Directions: Students place ropes in the center of play making sure that the ropes intersect one another. Students randomly turn over 2 cards and place one card in each circle. They then must find the card that will go in the intersection of the circles and place that card in the Intersection.

Example: If "large" and "red" are drawn, they must find the third card that says "large and red" and place it in the intersection of the ropes. The students then try to place the Apples in their appropriate places in the circle.

Activity 6: Game Board

Materials needed: Game board, markers, set of word dice or number the

Directions: The game board has a variety of Apples hanging on a tree in a game-like pattern.

Variation 1 The students will roll the three dice and make a logic sentence, and move to the Apple described by the sentence.

Example: if the words "red", "worm", or "or" are rolled, the player may move to a red Apple with a worm, or a red apple, or an apple with a worm.

Variation 2: The student rolls a die with the numbers 1, 2, and 3 and moves to an apple which has that number of attributes different.

Variation 3: Use cards instead of a die. The cards are shuffled and placed face down on the table. If a card with one line is drawn, the student moves to an apple that has only one attribute difference from the first. If a three is turned over, the player must find an apple with three different attributes.

Activity 7: Let's go Floating

Materials needed: Game board (file folder with large lake in the center and leaves floating on the lake), set of miniature apples, 3 dice (2 with attribute words and one with the words "and", "or", "not"), leaves (gives the apples a place to rest)

Directions: Three dice. Roll the dice and place an apple on the lake, which matches the description on the die.

One die: Randomly places one apple on a leaf on the lake. Roll the die and place another apple on the lake that has a different attribute from the first one. Example: If an "I" is rolled, the player places an apple on the lake that is one different from the first apple.

Cards: This is similar to the die, except the cards have 1, 2 or 3 lines drawn on them. The cards are shuffled and placed face down. If an 'I' is turned over an apple with one attribute different is placed on the lake.

The winner of the game is the last player to place an apple on the lake. If a player can't make a play, s/he looses a turn.

Activity 8: It's Getting Cold Let's Go Home!

Materials needed: Same as for activity 7

Directions: All of the apples are placed on the lake - activity 7 is reversed. This time place all the apples on the lake, and remove then according to the dice or cards.

Example: One apple is randomly taken off the lake, one die is rolled (or one card drawn) and the number "2" comes up. An apple must be taken off the lake, which has two attributes different from the one randomly taken off.

Example: *Three Die: If "red worm or" are rolled, a red apple with a worm can be taken off, or a red apple, or an apple with a worm.*
The winner is the last player to remove an apple.

These are a few activities that can be used with your apple.

Additional attribute model activities are in Appendix D.

Be creative and design activities that can be used in your classrooms.

WHY PATTERN?

- Lay the foundation for later development
- Later topics depend on the ability to discover and use patterns
- Future work is built on the discovery of basic patterns.
- Multiplication, rhyming words. Sentences, series, sequencing, etc.

Engaging students in a search for patterns is pedagogically rewarding and a pattern for a higher developmental level.

Patterns may be presented to the eye or ear - simple to complex. Suggested development of patterning:

- Copy a pattern Give the student a pattern '*A**A**A
- Have the student copy the pattern
- Identify and extend a pattern
- Give the student a pattern **A**B'*A
- Extend a pattern

As student get older patterns become more complicated, however, the first patterns should be strong and uncomplicated.

Activity 1: Patterning using body movements

Materials needed*: student and teacher*

Directions*: The teacher claps, stamps, makes some sort of a movement and the student repeats the movement. This is a good activity when you are trying to gain the attention of your class. Instead of using verbal commands, begin to do patterns with the body such as touch your head, wave your hand, place you hand above your head, touch your elbow as you begin to do the body movements, a few students will follow and gradually the whole class will be doing the patterning. Once you have their attention you can proceed with what you wanted to do.*

> ***Activity 2: Music*** *patterns abound in music*
>
> ***Materials needed****: rhythm band instruments (if they are not available use coffees can drums and pieces of wood for sticks)*
>
> ***Directions****: Teacher demonstrates a pattern, students follow the pattern*
>
> *Vary the pattern (tap, tap, hand at side, tap)*
>
> *What is the pattern in songs?*
>
> *Ask the students if they can repeat the pattern 3 times.*
>
> *Develop a rhythm band song or let them play their instrument to one of their favorite songs.*

Extension: Have the students work in pairs - one makes a pattern, the other repeats the pattern

> ***Activity 3: Patterning using manipulatives.***
>
> ***Materials needed****: beads, macaroni, pegboards, Cuisenaire rods, pattern blocks, paper chains, or any material that will make a pattern. (To follow the Teddy Bear theme, Teddy Grahams can be used.)*
>
> ***Directions****: Make a pattern and have the students repeat the pattern.*
>
> *(After students are familiar with pattern, they can be given a patterning guide to follow. They choose a pattern card, and follow the pattern. This can also be used for extending the pattern.*
>
> *For example a bead pattern showing 5 beads can be given and the student is a, repeat the pattern for the length of the bead string.*
>
> *This can be repeated with any of the manipulatives available.*

Learning Center Activities for Patterning:

Materials needed: stamp pads and stamp -these can be commercial stamps or home made stamps.), papers for stamp patterns.

Procedure: Have the students make repeating stamp patterns. Once they have made their own individual pattern. They share it with another student and the other student sees if they can repeat the identical pattern.

Commercial Materials for Patterning:

Commercial materials, which are available with accompanying books, are:

Pattern Blocks
Cuisenaire Rods
Tangram Puzzles

Homemade Materials for Patterning

There are many materials that we can find in our homes that can be used for patterning. The difficulty of the pattern will depend on the materials. Materials such as beans, pom poms. macaroni, spaghetti, rocks; etc. can be used for patterning activities. Students can either discover their own patterns, or repeat patterns described by the teacher.

Once students have had lots of experience with sorting, classifying, patterning, and ordering, they are ready to begin their adventures into the land of basic facts.

PATTERNING ACTIVITES FOR OLDER STUDENTS

All mathematics has a pattern. Sometimes it is difficult for younger students to see the pattern in addition and subtraction or multiplication and division. Each has its own unique pattern. If students can discover the number pattern, automaticity with the operations will be much easier.

Older students can use attribute models, which reinforces attributes and the where the attributes are not obvious. The models need to be made so they require thinking. Older students can also be asked to find patterns within the classroom; patterns on the playground; and patterns in the community.

Other suggested patterning activities for older students include:

1. Finding a pattern in a sequence of whole numbers and extend the sequence.

2. Extending patterns represented in tables or as ordered pairs and propose a rule to describe the relationship.

3. Identifying and graphing points in the coordinate plan and describe the results.

Activity 1: Growing Patterns

Materials needed: *inch cubes, cm cubes, triangles (equilateral)*

Directions: *Use the cubes to build a stair. Continue building stairs until you have 7 steps. How many cubes were required?*

Use the cubes and make the smallest rectangle (a square will not be considered a rectangle for this activity) Build the next largest rectangle. What is the pattern for the number of blocks required to build the next rectangle?

Use the cubes and build a smallest square. Continue building the next smallest s etc. What type of numbers are you building?

Use the triangles to build triangular numbers. List all of the triangular numbers that you can.

Activity 2: Number patterns.

<u>Materials needed</u>: sheets of paper with number patterns, pencil

<u>Directions</u>: Write several number patterns on a sheet of paper. Have the students the patterns in each sequence.

    2,4,6,8.10 ....

    1, 3, 4, 7, 11, 18.......

    1. 5, 9. 13, 17........

    2, 5,11,23,47.......

    What is a Fibonacci Number?

    1 1 2 3 5 8 13 21

    To discover the pattern, you must add the numbers together

    I.e. 1 + 1 = 2. 1 + 2 = 3, 2 + 3 = 5, etc.

Problem: What do we have around us that could be considered a Fibonacci Number?

How many bumps on the top of an apple?

This is a good way to get students involved in looking for patterns in real life situations.

CALCULATOR ACTIVITIES WITH PATTERNING

Activity 1: Calculator

<u>Materials</u>: Calculator

<u>Directions</u>: Press the "1 +* what number do you have?

<u>Problem</u>: How many 2's do you have to touch to get the number 36 to display?

    Pick any 1-digit number

    Multiply x 9 Record your Answer then clear your calculator

    Multiply by 99; Record your Answer then clear your calculator

    Multiply by 999; Record your Answer then clear your calculator

    Multiply by 99999. Record your Answer then clear your calculator

    What did you observe?

    Try it with other numbers such as 2. 3. Etc.

> *Activity 2: Calculator*
>
> <u>Materials needed</u>: calculator
>
> <u>Directions</u>: *You have a set of cards that start with 7, and go to 113. The first few cards say 8. 15. 22. How many cards are in the stack?*
>
> *Although there are many pattern activities, only a few have been mentioned for older students. As we go along during the rest of the quarter, we will be talking about other number patterns.*

THE CHILD'S YESTERDAY MAY HAVE BEEN RICH AND REWARDING AND FILLED WITH SUCCESS, OR THEIR YESTERDAY MAY HAVE BEEN THREADBARE, NEGLECTED AND DAMAGING TO THEIR EGO.

It is up to you as the teacher to make sure that the child's tomorrow is rich and rewarding.

Chapter 7

GEOMETRY

Formulas
Formulas
Formulas

What is the Formula for the area of a cube?

Is Geometry more than a series of formulas and proofs to be learned? Is there a place for Geometry in our everyday lives? Your answers to these questions may determine how you teach geometry in your classroom.

Many teachers see geometry as set of proofs and formulas, which must be solved in order to obtain the correct answer. Some of the teachers, who see geometry in this way, have developed an anxiety towards the concept. The mere mention of the word geometry can almost bring on a panic attack.

Prior to the 1970's those who took geometry had an interest in mathematics or in a career, which required geometry. Many teachers and parents never took geometry; therefore, they have a limited knowledge of the concept. Geometry was a subject, not part of their lives.

During the 1970's it was realized that geometry was an integral part of our everyday life. This caused educators to take another look at the concept and methods used to introduce it in a classroom environment. The conclusion was that geometry is present in our everyday life, and should become part of what we teach. Since this is the case, we need to find ways to introduce geometry as an extension of our everyday lives. We need to find ways to relate it to mathematics so students will understand it from Kindergarten through 12th grade. We need to teach the Geometry chapter in the book rather than skip it altogether – something that happens in many classrooms.

Geometry is a good way to begin a school year. The teacher can assess the student's mathematical ability through the multitude of skills required for geometry. This gives the students a break from the tedious review of last year's work, which they face at the beginning of each school year, and gives them a fresh look at mathematics.

Many teachers were not required to take Geometry in school; therefore, they have limited ability in teaching geometry. They were not exposed to this interesting world. Their memories are painful or non-existent and they tend to avoid the word and the operation in their own classrooms. In a survey of future teachers it was found that geometry was one of the areas in which they developed an anxiety toward mathematics. It has been stated that students have difficulty with geometry because they lack the spatial skills necessary to successfully understand geometry.

The 1980's ushered a new interest in geometry. Interest grew as we gained insight into how children learn an increased awareness of mathematical patterns and interrelationships found in our physical environment, and more clarity about how geometric and numerical concepts interrelate. The calculator and microcomputer have helped place more emphasis on geometry as an important component of elementary school mathematics.

Geometry is an important part of the everyday K-4 curriculum. Students at this age are naturally interested in geometry and find it intriguing and motivating. It is an extension of their preschool learning and their knowledge of geometric shapes and being able to manipulate them into forts, hideouts, etc. They are naturally curious and students at these grade levels have a means of strengthening their experiences through the teachers' guidance.

Geometry helps students build spatial understanding which will help them interpret the world around them. This understanding will enable them to develop insights and intuitions about two- and three dimensional shapes. Students who develop a strong sense in spatial relationships and who master the language are better prepared to be successful in higher level thinking skills required in more advanced mathematics.

Freudenthal in 1973 stated: "*Geometry is grasping space ... that space in which the child lives, breathes and moves. The space that the child must learn to know, explore, conquer, in order to live, breathe and move better in it.*'

Geometry helps students make sense out of their world. Through use of concrete geometric models, they discover relationships and develop a spatial sense of objects by drawing, measuring, visualizing, comparing, transforming, and classifying. This helps them build a knowledge bank by applying personal meaning to the abstract ideas. Definitions and formulas become meaningful relationships among figures they can discuss ideas, conjecture and test hypotheses – skills that precede formal operations. If presented in a manner which is non-threatening, it can provide an exciting and productive curriculum for middle school students. At this level geometry should focus on investigating and using geometric ideas and relationships rather than on memorizing definitions and formulas.

The new NCTM Standards stated that Mathematical instruction programs should include attention to geometry and spatial sense so that all students:

- Analyze characteristics and properties of two and three-dimensional geometric objects;
- Select and use different representational systems, including coordinate geometry and graph theory;
- Recognize the usefulness of transformations and symmetry in analyzing mathematical situations;
- Use visualization and spatial reasoning to solve problems both within and outside of mathematics.

The standards state that much of geometry should be learned through activity, with physical models, drawings, and dynamic software as learning tools. The eventual goal is for students to systematically study geometric shapes and structure and increasingly use formal reasoning and proof in their study.

GEOMETRY IS SEQUENTIAL

Geometry is sequential in nature. If it is not taught through this sequential development, frustration may occur. Students need to have hands-on experiences in order to develop geometry, which may mean teachers have to deviate from what is presented in the textbooks. If the prerequisites are not met, the concept becomes difficult and confusing.

Students go through different developmental stages in geometry. Whereas the textbook may determine the content of geometry, the stage will effect what the students are capable of learning. The different developmental stages are:

Before 4 years of age

- acquire topological concepts
- open/closed, inside/ outside /boundary, separation/ connectedness, order and proximity. Students are constantly changing views of familiar objects.

Age 4-7

- Begin to develop some Euclidean concepts. Euclidean studies size, shape, direction and angle.
- Perpendicularity and parallelism are important. They begin to see objects in relation to other objects.
- As they develop they begin to discriminate between circles and ellipses, triangles, rectangles, and parallelograms.
- Begin to recognize there is structure and order in space thus leading to the development of left and right.
- Crucial developments take place in children's perceptions and abilities. Children develop the ability to construct and conserve length and area.
- Children become aware that horizontal and vertical references can be simultaneous. They begin to investigate size and shape, and compare their relationship.

Age 9-11

- Begin to get dissatisfied with flat interpretations of space, and begin to fill in flat spaces. Mental capacities are maturing, and they are becoming more aware of details and relationships, and begin to develop geometric concepts, like: a square is a rectangle is a parallelogram.

There are two types of geometry - **Topological and Euclidean**. Children begin at the topological stage and move into the Euclidean stage.

During the topological stage students need to be encouraged to build with all types of materials (Cuisenaire rods, unfix cubes, pattern block, attribute blocks, geometric solids, geoblocks, etc). In order to help them develop the language, i.e. who is closest to, farthest from etc., the teacher needs to provide directions to help them focus on their activities.

Students move to the Euclidean stage is not sudden. It may occur over a two-year period. A student from 4-6 years old can recognize and name common shapes. (They may confuse a square and a rhombus or a rectangle and a parallelogram.) They begin to learn the names and properties of the common shapes and can name shapes within the room.

Piaget claims learning of shapes requires two actions:
1. physical handling of shape (running their fingers over the shape)
2. visual perception of shape itself

(visual perception can only occur when students have the shapes in their environment and are able to handle the shape and abstract the characteristics).

Children use proximity (nearness and farness), separation (separating the parts from the whole), order (sequence of events) and enclosure (inside or outside a curve and the concept of between). By the age of 8 a child is ready to consider Euclidean concepts related to rigid shapes, distance, straight lines and angles. Thus geometry at preschool and primary grades should stress topological ideas and should stress familiarity with objects from the three dimensional world. Only after the knowledge of the three dimensional world has been acquired should emphasis be placed on the two dimensional world. Research has shown that some Euclidean concepts are developed before topological concepts of order, equivalence and continuity.

We live in a three-dimensional world, and the child's experience with the world should be 3-dimensional. One teacher started her geometry unit by having her class take a Polaroid camera out into the community and take pictures of different geometric shapes. The class then made a scrapbook of the geometric shapes around the community.

When do children begin to learn geometry?

A child, at an early age, begins to notice and compare attributes such as shape, color, and length of objects. Much can be learned about geometry by observation. Pattern blocks, cuisenaire rods, tangrams, geometric solids, geoblocks, and other geometric shapes help children begin to analyze shapes by sorting and classifying the various materials. Young children explore and abstract meaning from objects through free play. Free play is also important for children of all ages. We learn by our own experience and classroom learning is no different.

Pattern Blocks: Pattern blocks are a commercial product which can be used throughout the grades. In the primary grades, students begin by exploring the blocks through free play. There are commercial pattern puzzles where students fill in the designs. The patterns could be placed in a learning center for free exploration. Primary aged students seem to prefer to make their own patterns with the blocks.

Question: _What do Primary students learn from playing with pattern blocks?_

- attributes of the shapes
- names of shapes
- the relationship between shapes
- patterning
- angles
- designing

Question: _What do intermediate students learn from playing with the pattern blocks?_

- all of the same things as primary aged students
- lines
- tessellations
- degrees in angles
- which shapes will fit together

Activities that can accompany pattern blocks:

Activity 1 – Shape Recognition

Math Concept: _Geometry - shape recognition_

Materials: _Pattern blocks,_

Vocabulary: _Pattern, square, rectangle, trapezoid, triangle, circle, parallelogram,_

Procedures: _After a period of free play, students sort the pattern blocks according to shape._

Game: "I'm Thinking of a Shape". The teacher says, "I am thinking of a shape that has 3 sides. Which shape am I thinking of?" After the teacher has given several shapes, the students can describe their shape.

Shape Comparison: Compare a square and a rectangle. How are they alike and how are they different. Divide the students into groups, give each group a sticky pad (either square or rectangle) and have them match the shape in the room with the shape of their sticky pad.

Add the trapezoid shape to the square and rectangle. Discuss how the three shapes are alike and different (All have 4 sides and 4 corners, the trapezoid's angles aren't the same size; the trapezoid also has 3 different sized sides, etc.) See if they can find any trapezoid shapes in the classroom.

Add the parallelogram to the set of shapes. The shape still has 4 sides, but now the corners (angles) are different. Discuss the shape of the parallelogram (kite). At this point in time it might be easier to let them use the word "kite" rather than parallelogram. They might have difficulty finding something in the room that looks like a kite. In this case, It might become a homework assignment -find a picture of something that is shaped like a kite.

After the students are familiar with the 4-sided shapes, add the circle and the triangle.

The Shape Walk: Take a shape walk around the school. Ask the students to look for the shapes they have been studying.

Shapes at Home: After each shape has been discussed, have the students see how many thing they can find at home for that shape. This is a good activity for homework

---

Activity 2: Pattern Blocks

Materials: *Pattern Block Shapes, papers, colored cutouts, Chart: Likeness and difference;*

Vocabulary: *pattern, square, rectangle, triangle, rhombus (kite), circle, trapezoid, parallelogram, hexagon, tessellation, and other words the teacher feels the students do not understand.*

Procedures: *Classification of Shapes: After a period of exploration, ask each group of students to classify the shapes in some manner, i.e. number of sides, same size side, same angle, etc. A discussion will follow as students describe their classifications. As students give the shape, number of sides, and number of angles, the teacher will record their response.*

The teacher will then ask the students to sort their shapes according to the categories listed on the board. A discussion will follow as students justify which names will go under each category.

Predicting and testing: Ask each group to predict the shapes, which will cover the piece of paper. Ask them to record their predictions. Once they have made their predictions, have them test their predictions. Were they correct? Why or Why not?

Tessellation Patterns: Using only the shapes that will cover the paper without leaving gaps, design a pattern. Trace the pattern or cut the pattern out of paper and glue the pattern down. Using the same shapes, design several patterns which will be used for a bulletin board. "Tessellations of the Day". Tessellations will be a new word for the students; therefore, the teacher will need to help them discover the meaning of the word. A discussion of the patterns on each card may help.

Mathematics, Social Studies and Art: Discuss the designs on Indian headbands. If possible bring in some Indian Headbands or pictures of them. Help the students notice the various geometric shapes used on the headbands. After they have discussed the designs, tell them. they are going to make their own headbands using geometric shapes. Give them several pieces of paper or a 3 x 5 index cards. Have them make the same pattern on each piece of paper or card (they might want to use small shapes to glue on the paper or they could use crayon to draw the design). They will need to connect enough paper or cards to go around their head. After they have completed all of the cards, punch holes in the card, tie the cards together with yarn and wear them as an Indian headband.

School Patterns: After the students have completed the activities, take another walk around the school and see if the students can discover other patterns.

Tile a floor. Bring in some pictures of patterns and ask the students to describe the patterns.

Activity 3: Tessellation's (3rd grade and up)

Tessellations are repeating patterns which do not gap or overlap. A tile floor is an example of a tessellation. Tessellation's can be simple or made more complex by cutting and reshaping to make an integrate patterns. Escher has given geometric shapes an artistic view. Escher has many books about tessellations.

Directions: *Give the student some geometric shapes (Xerox some shapes and let the students cut them out. Have them make tessellation patterns. They can then color in the pattern. For more advanced students, give them a 3 x 5 Index card. Have them cut the card and slide the cut piece to the opposite side. Tape the piece on to original piece.*

Trace around the pieces and make an artistic design.

The above activities are used to help the students gain knowledge of geometric shapes. In their activities they will learn about the lines and angles, discover what angles can be placed together to make a straight line, what angles will make a circle. etc. The characteristics they discover at this level will enable them to use geometry in many different ways. Geometry in preschool and primary grades should stress topological ideas, and should stress familiarity with objects from the three-dimensional world.

QUILTS

Geometric shapes and patterns blocks can be used for students to make their own quilts. The use of quilts in all grades is popular at this time. Many books have been written about quilts, and there is probably one that would be of interest to your students. There are even a few books that relate more to the older student. Students of any age can work on quilts. Quilts can be made of paper (regular, tissue, wrapping paper, etc.) or cloth. They can be colored or painted. They can actually be quilted by the students, or they can become a bulletin board.

> ### *Activity 4: Design Block quilts.*
>
> **Materials**: *Stories about quilts (The Josefina Story guilt by Eleanor Coerr), quilt shapes (pattern block pieces cut from scraps of material, or discarded wrapping paper, or construction paper), mat for the pieces, glue, real quilt.*
>
> **Procedures**: *Bring in a quilt if possible and discuss how the quilt was made of various shapes and the shapes are put together to make a large pattern. Read one of the many stories that has been written about a quilt, tell the students they are going to make a classroom quilt using the shapes. Each child will make one square (a piece of paper cut to a size such as 6 x 6 or larger) to place on the quilt. (The student's square should be used, even if it is not perfect. It is their square.) After each piece has been made, the students help the teacher put the quilt together. They must decide the color for the outside, and the strips in between.*

Extensions:

Create a learning center with various shapes and patterns. The students will be free to create other patterns.

Classroom Quilt: Use material to make a quilt. Give each student a square of white material and have him or her draw a design on the quilt. The squares are sewn together to make a large class quilt. (It is possible that a parent might know how to make quilts and would be willing to help with this project.)

Square Corners: Give the students an index card. Talk about the corner of the index card. Have them find objects in the classroom that have square corners like the index card. This is a good Introduction to square corners (90^0 angles), and it helps them become aware of the differences in angles.

Question: *Does the trapezoid have square corners?*

TANGRAMS

A tangram is an ancient Chinese puzzle. The seven pieces of puzzle can be put together to form a square. Tangrams are a commercial produced, however, the die cut machines have a tangram die cut, and a tangram puzzle can be cut out of a square piece of paper.

There are two books which can be used with the tangrams. The Tangram Magician by Ernst is an excellent book for younger children. It also contains a set of self-adhesive tangrams which the students can use to make different shapes. Grandfather Tangs Story is an excellent book for older children.

> **Activity 5: Grandfather Tang's Story**
>
> **Directions**: Read the book <u>Grandfather Tangs Story</u> by Ann Tompert or <u>The Tangram Magician</u> by Ernst. As you read the story make the shapes on the overhead.
>
> After the story has been read, let the children make their own animals. This can be expanded into a language lesson by having the children write stories about their own animals.

The above activities are used to help the students gain knowledge of geometric shapes. In these activities they will learn about the lines and angles, discover what angles can be placed together to make a straight line, what angles will make a circle, etc. The characteristics they discover at this level will enable them to use geometry in may different ways. You will also notice the emphasis has not been on the name of the shape, but rather the characteristics.

GEOBOARD GEOMETRY

Geoboards are a manipulative which can be used at any grade level. Many teachers are reluctant to use geoboards due to the use of rubber bands or geobands. If the classroom rules are established regarding rubber band, the students will usually follow the rules. Geoboards help students discover various geometric attributes. All students need to begin with a period of free play. As they "play" with the boards, they will discover they can make various geometric shapes and practice making angles.

Young children begin with making designs with rubber bands – and usually see how many they can put on one geoboard. This is probably also true of older students; however, the older ones usually produce designs and shapes and seem to have a direction for their free play. Geoboards can be used as introductory activities for area and perimeter.

(Geoboards they are easily made with boards and nails. To get the nails even, tape graph paper on the board and use the intersections for the nails.)

> **Activity 6: Play**
>
> **Directions**: Place the geoboards and rubber bands in a learning center for experimentation.

> **Activity 7: Line Segment Game**
>
> **Directions**. Two players the first player makes a line segment. The second player makes another line segment from either end of the first. Play continues until no more line segments can be made
>
> **RULE**: the lines cannot cross.

Activity 8: Make a figure

Directions: Make an open curve. Make a closed curve. Make a 3-sided figure. (Show the different three sided figures and discuss the likenesses and differences this opens the opportunity to discuss the different types of triangles.) Turn the board 1/4 turn. Does it look different? Turn it another 1/4 turn. Make a 4-sided figure. Make the largest 4-sided figure smallest.

Activity 9: Do as I do.

Directions: One player makes a design, and second player must copy the design. Both players then make a design that is symmetrical to the first design.

Activity 10: Symmetry

Vocabulary: Symmetry.

Directions: Make a design on the geoboard - have the next person make a shape that is symmetrical to the first shape.

Activity 11: Introduction to area *(This can be introduced at an early level.)*

Vocabulary: area

Directions: Explain the meaning of a unit on the geoboard.

1. Construct a square with an area of one unit
2. Construct a square with an area of 4
3. Construct a square with an area of 6
4. Construct a square with an area of 1, a square with an area of 3, and a square with an area of 5. If any are impossible, state why.
5. Construct the largest possible square on the geoboard. What is the area of the square?

Activity 12: Perimeter

Vocabulary: perimeter

Directions: Ask the following questions:

1. What is the perimeter of a square with an area of 1?
2. What is the perimeter of a square with an area of 4?

What is the perimeter of the largest square you made on the geoboard?

Activity 13: Rolling for Rectangles/ Perimeter/Area

Materials needed: 2 die centimeter graph paper, colored pencils

Directions: Roll 2 dice. One die represents the length and the other the width of the rectangle. Mark the square or rectangle indicated by the die, i.e. 6 x 3. If you are rolling the perimeter, you will outline the shape. If you are rolling for the area, color in the squares. Repeat 3 to 5 times.

Question: When does the perimeter become larger than the area?

Activity 14: Rectangles Come in Many Sizes

Materials needed: centimeter paper

Directions: Give each student a number and have them make as many arrangements of that number as possible. For example if the number is 12 - they can make a 1 x 12, 2 x 6, or a 3 x 4 arrangement.

Activity 15: Move the Shapes

Vocabulary: Rhombus, Parallelogram

Directions: Make a square with an area of 4. Move the rubber band on the top right side of the square one peg to the right. Move the rubber band on the top left side of the square one peg to the right. What shape have you made?

Make a rectangle 2 pegs high and 3 pegs wide. Move the rubber band on the top right side of the rectangle one peg to the right. Move the rubber band on the top left side one peg to the right. What shape have you made?

Extension using other than geoboards

Math Concept: rotations. slides, lines, angles, shapes

Materials: Strips of oak tag, brads

Procedures: Square: Give each student four equal sized strips of oak tag and four brads. Punch a hole at each end of the strip. Use brads to fasten the strips together to make a square. Rectangle: Give each student four strips of oak tag 2 long and 2 short. Punch holes in each end of the strip and fasten together with brads.

Students can now experiment as they change the shape of a square. They can push on the comer and make a rhombus or diamond. They can use the rectangle to make parallelograms. These experiences with the square and rectangle will also help them later when they are determining the area and perimeter of the shape.

Activity 16: Other Shapes

Vocabulary: Circle, Triangle

Directions: Ask the student to make a circle on the geoboard. Why is it impossible to make a circle on the geoboard?

Ask each student to make a triangle on his or her geoboard. After they have each made their triangle, ask them to compare their triangles to others in the group.

The teacher will help the students classify the triangles according to the type (equilateral - all sides congruent; scalene - no sides congruent; isosceles - only two sides congruent).

The group decides what their triangles have in common. After each group has made their decisions, they report to the rest of the class. The teacher then names the types of triangles. The teacher gives each student a protractor to measure the angles. As they give the measurement of each angle, the teacher records the measurements in three columns - right angles; obtuse angles (an angle greater than a right angle); acute angles (all angles smaller than right angles);.

Ask the students to rotate their triangles one-quarter turn. Discuss the differences in the triangles

Activity 17: Area of Triangles

Vocabulary: Area, types of triangles,

Directions: Make a right triangle on your geoboard. What is the area of your right triangle? (The triangle cuts between pegs and it is not as easy to count the number of spaces) Use another rubber band and make a square around your triangle.

- What is the area of the square?
- What is the relationship of the triangle to the square?
- What is the area of the right triangle?
- Make an acute triangle on your geoboard.
- What is the area of the triangle?
- How can you find the area?
- (Make two rectangles around your triangle. Find the area of the rectangle and then the area of the triangle.)
- Make an obtuse triangle on your geoboard.
- What is the area of the obtuse triangle?
- Is the same method used for all triangles?
- What is the formula for finding the area of a triangle?
- Why do we have to multiply by 1/2?

Extension: Write a note to your friend who does not understand how to find the area of a triangle. Give your friend some hints as to ways to find the area.

> **Activity 18: Area of Polygons**
>
> **Vocabulary**: Polygon
>
> **Directions**: Make a polygon on your geoboard.
>
> If you move the rubber band one peg to the right on both sides of the top, will it change the area?
>
> Find the area of the polygon. You should be able to use the method we just used for finding the area of the triangles. Explain how you found the area of your polygon.
>
> Make up your own polygon figure. Give it to someone else in your group and have them determine the area of the figure.

GEOMETRIC SOLIDS

Geometric solids are another excellent manipulative that can be used throughout the grades. (See Appendix B)

The students in the early primary grades use them for building and begin to abstract attributes of the models. The older students can use them to help them discover area, perimeter, surface area, volume and vocabulary such as lines, vertex, etc.

Older students can also use geoblocks to reinforce the various words: vertex, edge, face, and surface area. Geoblock jackets can be made to help students fit the jackets on the various shapes. The first geoblock jackets should be made with graph paper. They should also be shapes that are square or rectangular. Once any fact becomes a triangle, it becomes more difficult to determine the surface area. Once students become proficient with shapes that are all rectangular, they can be introduced to irregular shapes.

Activity: Have the students bring in boxes, e.g. cereal boxes, to use in determining the surface area. The students will need to cut it so it will open flat, but yet be in one piece. Help the students determine how to find the surface area. This will lead them to discover the formula for a surface area.

ANGLES

Angles are an integral part of geometry. Students need to have a variety of experiences with angles. We start with a simple straight line.

> **Activity 19: Degrees in a straight line**
>
> Have the students place two dots on a piece of paper and connect them using a ruler. This is called 180 degrees. Help them understand all straight lines have the same number of degrees.

> ☿ *Give the students a straight strip of paper. Ask them the number of degrees in the straight line. Ask them to cut the strip in half and arrange it to make a corner. If the straight line is 180 degrees, how many degrees in the new angle?*
>
> ☿ *Ask them to cut each of the two pieces in half and make the four pieces into a square. Lead them to discover they now have 4 corners. each corner is 90 degrees, so the degrees of a square are 360 degrees.*
>
> ☿ *Give the students another strip of paper. Have them place it from corner to corner, diagonally across the square. What shape does that make? If a square is 360 degrees, how many degrees if in a triangle? (180 degrees)*

Activity 20: Angles in Triangle

Directions: Give each student two 3 x 5 index cards and a ruler. Ask them to construct and cut out a triangle. Make a large dot on each corner of the triangle. Have them use the first triangle as a pattern to make a second triangle.

Discover the angles in a triangle:

- Draw a triangle on a 3 x 5 card.
- Cut out the triangle and compare it with the other triangles.
- Group the triangles according to acute, obtuse, or right.
- Use a ruler to measure the sides of the triangles.
- Make a list on the board of the dimensions and list them according to:
- All sides equal (equilateral)
- All sides have different measurements (scalene)
- Legs are equal, i.e. 4, 4, 3 (isosceles)
- Trace a second triangle on another 3 x 5 card and cut it out.
- Place a dot on each corner of the triangle.
- Cut or tear off all three corners. Put them together in a straight line.
- How many degrees in a straight line? (180°)
- How many degrees in a triangle? (180°)

Find the missing angles for the following triangles? List the type of triangle.

    30, 50 _____   _____triangle

    90, 45, ____   _____triangle

    60, 60 _____   _____triangle

The students will compare their triangle with their neighbor's and then show it to the rest of the class. A discussion of the likeness and differences of the triangles will follow. To further discover the likeness and differences, have the students use a ruler to measure the sides of their triangle. As the students give the measurements, the teacher will list the measurements under the appropriate columns; i.e. if a student says all sides are 4, 4, and 4 (equal), the teacher would list the measurements under a column called equilateral triangles. If all of the sides have different measurements (2, 3, and 4), the measurements will be listed under scalene and if the legs are equals (4.4, 3) the measurements will be listed under isosceles.

The students will need to take their second triangle and label the angles. The point should be labeled "B" which is the vertex of the triangle. (If the students do not already know the word "vertex", introduce it at this time.) Label the other two angles A and C. After the angles are labeled, the students will cut or tear off angle A and C and place all of the dots together. The three angles should make a straight line or 180°. The teacher can write <A + <B + <C = 180°

AREA OF TRIANGLES

Materials needed: Geoboard, geobands

<u>Make a triangle on your geoboard</u>
- Make a square around the triangle. What is the relationship of the square to the triangle?
- What is the area of the square?
- What is the area of the triangle?

<u>Make an acute triangle on your geoboard.</u>
- What is the area of this triangle?
- Make two rectangles around the triangle. Find the area of the rectangle and then the area of the triangle.

<u>Make an obtuse triangle on your geoboard.</u>
- Can you use the same technique to discover the area of the triangle?
- What is the formula for finding the area of a square or rectangle?
- Why do we say the formula for a triangle is ½ BH?

Angles in other shapes:

Number of degrees in a circle? _____

Number of degrees in a square? _____

Number of degrees in a rectangle? _____

<u>Make a polygon on the geoboard.</u>
- What is the area of the polygon?
- Cut out a polygon that is the same as the one you made on the geoboard.
- Fold one side straight.

- ℧ Cut off the side and move it to the other side of the polygon.
- ℧ What shape did you make?

Activity 21: Degrees in a circle

Vocabulary: *corner, degrees*

Materials: *index card, crayon or pencil, scissors*

Directions: *Color each corner of the index card. Cut off the corners (make sure they are large enough to handle). Place the colored corners together. Place a penny on the place where the four colored corners come together. Trace around the penny and remove it. What shape do you have? How many degrees in each colored corner? How many degrees all together? Is there a relationship between the number of degrees in a circle and in a square?*

Activity 22 Pi π

Vocabulary: diameter, circumference, and radius

Materials: Several cans and containers

Directions: Use a piece of paper or adding matching tape to measure around the outside of the can.

Circumference of a Circle - Many of us learned the formula for the circumference of a circle. Circumference = π x diameter

Circumference = 2 x π x radius or $2\pi r$

We never stop to question what the formula needs. In order to help students, you need to help them visualize the formula.

Extension Activity: Circle (lid, container, etc.)

- Use a piece of paper to measure the circumference of the object.
- Use a second piece of paper to measure the diameter of the object.
- Cut a second piece of paper the exact measurement of the circumference. How many pieces the size of the diameter can you cut from the circumference?
- Use a tape measure to measure the circumference and diameter of the containers.
- Use your calculator to divide the circumference by the diameter. What is your answer?

Activity 23: PI

Vocabulary: area, circumference, radius, and diameter

Materials needed: round objects, rulers string, measuring tape

Directions:

- Give each group 2 pieces of string.
- Use one piece to measure the distance around the can. Mark the measurement.
- Use the second piece of string to measure to diameter of the can or object.

Question: What is the comparison between circumference and the diameter?

- Make a mark on a round object.
- Mark a starting point on the table.
- Put the mark on the object.
- Roll the object until the mark touches the table again.
- Measure the distance.
- Draw a line for that measurement.
- Measure the diameter of the object.
- Draw a line to represent that measurement.

Questions: How do the two measurements compare?

- **How many diameters can be made from the circumference?**

Area of the circle: $A = \pi r^2$.

Surface Area of a Cylinder

The surface area is the total outside surface of the cylinder.
The distance around the can is determined by $2\pi rh$.
Add to that area the area of both ends = $2\pi r^2$.

Formula: $2\pi r^2 + 2\pi rh$

Activity 24: Area and Perimeter of paper samples

Directions: Use centimeter paper to make several different squares and rectangles. To determine the area of the object, see how many centimeter cubes will fit on the paper. If you have a 2 x 2 square, you will be able to fit four one-centimeter cubes on the paper to determine the area. Can you find an easy way of determining the area of the shapes? What formula would you use to find the area?

To determine the perimeter. How many one-centimeter cubes will fit around the outside of the shape? Can you find an easy way of determining the perimeter of the shapes? What formula would you use to find the perimeter?

Activity 25: Volume

Directions: Make several boxes! Note: The first boxes should have the square showing, and then move to boxes without squares.

Have the students cut out boxes A. B. C, D. E, and tape them together. Which box will hold the most cubes? Which box will hold the least cubes?

They will estimate how many cubes it will take to fill each box.

Estimate Answer

A _____
B _____
C _____
D _____
E _____

Give the students graph paper and ask them to make boxes with the following measurements:

Length Width Height Guess Answer
  2   3   3
  5   2   3
  7   1   4

After completion of this activity, see if you can determine the number of cubes the following boxes would hold?

Box   Length Width Height Guess
          12   5   2
           6   25   2
         102   5   4

Writing: How would you determine the volume of each?

To determine the volume, you want to know how many centimeter cubes will fit into the box.

VOLUME

Question: How does volume differ from area and perimeter?

Use various sized boxes to help students discover the meaning of volume. Small centimeter cubes can be used to fill the box.

Formula: length x width x height

Writing: What is the difference between areas? perimeter and volume?

Hands-on experiences with area, perimeter, volume, pi, and angles will enhance the student's geometric knowledge base and give them a point of visualization which will enable them to discover and solve the formulas for geometry. Once they are able to see the reason for the formula, it is no longer a foreign language. It relates to the old Chinese Proverb - what I hear, I forget; what I see I remember, and what I do, I understand! Students will understand and be able to commit the formulas to their long term memory.

Geometry is a constant in mathematics. Although we think of it as shapes, and finding areas and perimeters, it is much more than that. The formulas which all of you know can be easily developed from the activities. We also see how geometry relates to other operations in mathematics. Rational numbers use geometric models to discover ratio, proportion, percentages and decimals. Multiplication facts must form either a square or a rectangle. Basic Facts relate to odd and even numbers – which relates to geometry. As we think through all mathematical ideas, we see their strong relationship to geometry.

Monster Geometry

Monster Books: A book about monster could be used as an introduction to this activity

Math Concept: geometric shapes, lines and angles, critical thinking, literacy

Materials: Construction paper, glue, scissors, tape, decorations

Procedures: Each student makes a geometric monster. The monster can be as large or small as you want it. It can be 3 D or 2 D. After the monster is completed, give the monster a name and write a riddle about the monster. (Do not let anyone see the name or hear your riddle.) Put the name of your monster in the name pile and the riddle in a riddle pile. After all of the monsters are completed, display them in some manner. The Monster riddles are handed out to the group - be sure no one gets their own riddle and the group tries to decide which monster is being described. Once the riddles are matched to the monsters, then the names are passed out and the names assigned to a monster.

Write a story about the monster.

Activity 26: Monster Activities:
Procedures: *Once the monsters are displayed, give the groups the following list of questions: (A certificate could be made to go with each of the categories, i.e. most shapes, tallest, shortest, etc.)*

- *Which monster used the most geometric shapes?*
- *Which monsters used the least number of geometric shapes?*
- *Which monster's hat has the most edges?*
- *Which monster's hat has the most corners?*
- *Which monster is the tallest?*
- Which monster is the shortest?
- *Which monster has the largest hat?*

Activity 27: Symmetry
Math Concept: *Symmetry*

Vocabulary*: Symmetrical*

Materials: *construction paper, scissors, and mirror, large letters of the alphabet (both small and capital), large numbers, shapes, paint*

Procedures: *Take one piece of construction paper and fold it in half. Hold the folded edge and cut a design. Open up the design and see what you have made. Discuss the likeness of the two sides of the design.*

Problem*: Folding the paper in half and cutting it can make what other shapes? Note: Students can make their own monthly calendar by cutting symmetrical designs that either go with an event during the month or a classroom theme. The shapes are placed around a blank calendar, and as the day arrives, the student who made the design with that number on it will place the appropriate day on the calendar.*

Some suggested shapes are: September*: leaves, trees, schools, bells;* October*: ghosts, brooms, pumpkins;* November*: turkey feathers, pilgrim hats, pilgrim faces;* December*: trees, Christmas Decorations, Hanukkah Decorations;* January: *snowflakes, bare trees, clouds* February*: valentines, cherries, top hats;* March: *shamrocks, Easter decorations (depending when Easter falls), kites;* April: *Easter Decorations. flowers, bunnies;* May*: flowers, sun*

Symmetrical shapes can also be made to go with a theme.

If a student has a birthday during that month, have them make a birthday cake to go on the day.

Art: Fold a piece of paper in half. Open it up and drop some different colors of paint on one side. After the paint has been applied, fold the sheet over, and carefully press the two sheets together. Open it up. Look at the design.

Problem: Which capital letters in the alphabet are symmetrical? Are any of the small letters symmetrical? If so, which ones? Are any of the numbers symmetrical? Is your name symmetrical?

Procedures: Give each student a mirror and the letters or numbers. Have them place the mirror at the cent of the letter or number, and see if the letter or number is symmetrical.

Variation: Give the students half of a letter or number and see if they can determine the letter or number by placing the mirror on the half.

Geometry is Fun!

Teachers can teach geometry and open up new horizons for their students. The students' future with geometry depends on the excitement and interest created by the teacher.

Chapter 8
MEASUREMENT

Our lives revolve around measurement. It is so much a part of our everyday life that we do not realize we are using it. Money, calendars, cooking, driving, etc. all use measurement. Think about how much you use measurement during the day.

Historically, the Early Egyptians, Babylonians, Greeks and Romans used body parts as units of measurement. They used foot measurements, finger measurements, hand measurements, etc. to tell others the size of things. A time came when body measurements were not enough - people's feet differed in size, their hands were not the same. Therefore, a new system was devised, leading to the birth of our standard measurements.

The United States uses the English measurement (inches. quarts, etc.), while the rest of the world uses metric measurement, (liter, kilometers, centimeters, etc.) Milne (1892) stated

> *The Metric System of weights and measures is used by most of the civilized nations of the world except the United States and Great Britain and some of her colonies. (p. 412)**

We could use the same statement today, except leave out the words 'some of her colonies". Change takes time, but we have been attempting this change since the early 19th century.

Measurement requires sequential activities just the same as all other mathematics. Children begin measurement before they enter school. Most of their experiences have been hands on experiences where they were free to explore and develop their own meaning. Once they enter school, they should be able to use their past experiences and expand their knowledge of measurement. Measurement activities should begin at the Kindergarten level and spiral as they move from grade to grade.

The 2000 NCTM Standard states that students should become proficient in the use of measurement tools, techniques and formulas in a range of situations. The standards state that measurement is an important part of the mathematics curriculum from prekindergarten through high school due to its practicality and use in everyday life.

The study of measurement gives students the opportunity to apply it through other number operations such as geometry, statistics, and functions, plus basic operations. Measurement helps us make connections within and between mathematics and in outside of mathematics in areas such as social studies, science, art, and physical education.

The Standards have given examples of what students should learn at different stages of development. Primary students should learn measurement such as counting, estimating, and using formulas. Their math learning should include many materials and techniques. It also states that tools such as rulers and analog clocks should be used at this age. Estimation techniques will help them better understand the process of measurement and the size of the unit.

Elementary and middle-grades should continue to use these techniques and develop new ones. In addition, they ought to begin to adapt their current tools and invent new techniques to find more-complicated measurements. They should develop the formulas for perimeter and area and at the middle grades formalize the techniques and formula for volume and surface area of objects like prisms and cylinders.

No matter what age is being taught, it is the teachers job to help the students "bridge the gap" between the objects and the formulas. All students should have many opportunities to estimate and compare sizes of units and gradually begin to use a benchmark for their comparison.

Students in grades 3-5 should have opportunities to use maps and make simple drawings. Students in grades 6-8 should extend their understanding to scaling to solve problems. Middle schools students should use benchmarks to estimate angle measures and should estimate derived measurement such as speed.

The expectations for Pre-K-2 are:

- *Recognize the attributes of length, volume, weight, area, and time*
- *Compare and order objects according to these attributes*
- *Understand how to measure using nonstandard and standard units*
- *Select an appropriate unit and tool for the attribute being measured.*

Measurement is used in all of mathematics. It is sometimes called the bridge between number and geometry and can teach many daily skills. Measurement also helps strengthen the students' knowledge of other important topics.

The expectations in Grades 3-5 are:

- *Identify, compare, and analyze attributes to two- and three dimensional shapes and develop vocabulary to describe the attributes*
- *Classify two-and three dimensional shapes according to their properties and develop definitions of classes of shapes such as triangles and pyramids*
- *Investigate, describe, and reason about the results of subdividing, combining, and transforming shapes*
- *Explore congruence and similarity;*
- *Make and test conjectures about geometric properties and relationships and develop logical arguments to justify conclusions*

Measurement requires the students to reason. In the middle grades the students will begin to investigate the geometric problems with increasing complexity. As the students move from one grade to another, they should begin to describe the properties of various geometric shapes, i.e. rectangle, triangle, pyramid, or prism. As their reasoning skills become more complex, they should be able to explore motion, location and orientation such as flips and turns to demonstrate two shapes are congruent. Understand both metric and customary systems of measurement

Grades 6-8

- *Understand relationships among units and convert from one unit to another within the same system*
- *Understand, select, and use units of appropriate size and type of angles, perimeter, area, surface area, and volume.*

As students move from grade to grade, they draw on their previous experiences. They will move from the informal to formal stages of measurement. All students should understand the measurable attributes of objects.

Measurement can bring fun into mathematics. All students enjoy measurement and need to be able to develop an understanding of it through the use of sequential activities. The concepts that are to be taught are not part of the "new new math", but following the way measurement was originally developed, or we might say "the old old math". The sequential steps needed to help students development the concept for measurement are:

Comparison: students compare objects simply by measuring one against the other.

Non Standard

Body- this is probably the most used of all types of measurement. We use some part of the body to do measurement, i.e. feet to measure the distance across the room.

Non-body - we take what ever we have available to determine the measurement - piece of string, paper clips, pencils, etc.

Standard - this is the first measurement where we are actually using a measurement device. In the standard measurement we are only interested in large measurements - inches, centimeters, liters. Etc.

Precision - we need the precise measurement to the smallest unit.

These four steps (Comparison, non standard, standard and precision) are sequential in development, and students need to begin their measurement activities at the comparison level no matter what their age.

Each measurement activity should begin at the comparison level and progress until it is at the developmental level of the students – irresponsive of their age or grade level.

There are different types of measurement. We have linear, area, perimeter, clock, money, etc. Each can be developed so all students understand them and can do operations with the concepts.

LINEAR MEASUREMENT

An easy way to show the different steps used in teaching measurement is to take an activity that everyone can do such as measuring height.

- **Comparison** - Stand back to back and determine who is tallest and shortest. Line up according to height – shortest to tallest.
- **Non Standard body** – a person could lie down on the floor and see how many feet long they are. Several people could use their feet to measure the person and then compare to see if there is a difference in the measurements.
- **Non-standard non body**- Use adding tape or string to measure your height. Use masking tape to attach the tape or string to the wall according to height.

NOTE: Students in the early primary grades usually use comparison and non-standard measurement. Older students will be able to use the total four-step process. It is Important for older students to experience the comparison and non-standard measurements as well as standard and precision.

- **Standard** - Use a tape measure or yardstick to determine the height in inches or centimeters. I.e., the measurements will be approximately 54 inches tall or 176 cm tall.
- **Precision** -Give a precise measurement of your height

    5 feet 4 1 / 2 inches, or 1 meter, 87 centimeters.

Activity 1: Linear Measurement

Vocabulary: length, width, hands span. height, feet

Directions: Spread out your hand and draw around it. Cut it out. Use the four-step process to complete your hand measurement.

Use your hand span and see if there is a relationship between the hand span and the height.

Other comparisons:

> width of hand (fingers closed)
> length of thumb
> length of foot
> size of shoes.
> Are there any relationship between the above and your height?

Activity 2: Piece of furniture in classroom

Materials needed: Light switch or other piece of furniture in the classroom

Directions: Use the four-step process to determine the height of the light switch.

Comparison: Stand next to the light switch and use body parts to determine the height of the light switch. Does the light switch measure against the same body part for all persons in the group?

Non-Standard Body - this step can be skipped since we used a body measurement for comparison

Non-Standard Non Body - use string or adding machine tape to measure the light switch

Standard Unit - use a ruler (either English or Metric) or yard / meter stick. First record only the number of rulers or yardsticks needed. Second be more precise and measure the inches or half inches.

Precision: Use the yardstick or tape measure and give a precise measurement.

Activity 3: Classroom Measurement

Comparison: Compare different classrooms the students are in during the day.

Non-standard Body: How many children fit across the length and width of the room?

Non-standard Non Body: How many unifex cubes, or chain lines, etc. are needed to measure the length and width of the room?

Standard - How many rulers or yardsticks are needed?

Precision - What is the exact measurement of the room?

The following activity will help students discover something about themselves (older students may be able to relate to this better than younger students).

Activity 3 – Bouncing Balls
(set of balls labeled with letters).

Comparison: Which ball is larger, smaller? Prediction: Which will bounce higher?

Non-standard body: Bounce the balls

Put your hand on the body part that indicates the height of the bounce.

Non-standard non-body put a piece of paper on the wall. Mark the height it reaches.

Standard - Place some butcher paper on the wall, which has been marked off in 6-inch intervals. The students make a mark on the height of the bounce then count the number of 6-inch marks and record their answer.

Precision The marks from the standard can be used to find the exact height.

Other media in the classroom that could be used: light switches, desks

Activity 4: Discover your body measurements

Materials needed: string, scissors,

Directions: Take a piece of string and measure your height. Use this piece of string to measure your arm span (from your nose to the tip of your finger). Continue using this piece of string to measure your head; waist; neck; wrist and thumb. What is the relationship of all of these? Can you write a formula, which states the number of thumbs that make a neck?

Activity 5: What is your Shape?

Materials needed: string, scissors,

Directions: Take a piece of string and measure your height. Take a second piece of string and measure your arm width. Compare the two strings

If they are equal you are a square, if they are not equal you are a rectangle.

Variation: Graph your height and arm span. Determine how many different types of shapes your group makes.

AREA MEASUREMENT

A good introduction to area is the book "The King's Foot" by Rolf Myler (Del Press, 1990). This book introduces the concept of standardized measurement. It tells about a king who wanted to give the queen a special birthday present - a bed. The difference in measurement leads to a problem in the present for the queen. The following activities can be used after reading the book.

Math Concepts: nonstandard measurement; statistics (mean, median, mode for older students; English measurement; metric measurement; graphing: area and perimeter)

Activity 1: Compare your feet

Materials needed: sheet of paper, pencil, scissors

Vocabulary: surface, area

Directions: Each member of the group will estimate the size of their feet. They will do this by finding a way to compare the size of their foot. They will line up according to the length of their foot.

Comparison

Materials needed: graph paper, pencil, scissors

Directions:

- Use graph paper and draw around your right foot. Cut the foot out.
- Use the same procedure for your left foot. Compare the size of each foot.
- Line up according to foot length.
- Line up according to the width of your foot. Have you changed places?
- Place the feet on the bulletin board according to length.

Non Standard Unit

Materials needed: foot or foot shape, unfix cubes, centimeter cubes, etc.

Directions: Find out how many unfix cubes and centimeter cubes are needed to cover the foot. You may need to estimate the parts of the cube etc. or you might want to cut the parts out and put them together to make a whole.

Standard Unit

Materials needed: Graph paper, foot shape, scissors

Directions: Determine the area of the foot in inches (write the area on front of the foot). Repeat the same exercise using centimeter paper. In both cases, use only the whole squares.

Precision

Materials needed: same materials as in previous activity

Directions: This time see if you can determine the exact area of the foot. You may need to cut the half or part squares, and place them together to make a whole square.

Graphing: Graphs the footprints for the whole class,

one with length and the other with area. Can you predict the order of the footprints?

Extension of activities:

Manipulatives: two different sized feet (the die cut machine comes with two different-sized feet, or a length measurement can be taken for the foot. One foot cut from inch graph paper, and the other foot cut from centimeter graph paper. If it is not available, have the students trace around their stocking foot (designate either left or right foot), and cut them out. The teacher can then pair the students together to measure different objects within the room.

Upper grade students: Have the students use their feet to measure their bed at home. They then transfer those measurements to inch or centimeter graph paper. They make their beds and determine the area of the bed. A good comparison is in the difference between the centimeter beds and the inch beds.

Group Activities:

Non standard measurement

Students will work in pairs for this exercise. Each student will measure the object and record the measurement.

Make a worksheet where the students measure different objects in the room. An example

Object	Student 1	Student 2
table	__feet	__feet
desk	__feet	__feet
height of light switch	__feet	__feet

Other objects will depend on what is available in your classroom.

Total Class Activity:

List the measurements for the various objects on the board. Make a comparison to the largest and the smallest foot.

Discussion: Why is there a difference when everyone used the same manipulative?

Have the students make a graph, using their feet.

Make a Bed: Since the story is about making a bed for the queen, let each student design their own bed, using the measurements from their footprint. Give them either centimeter or inch graph paper. This will give them an opportunity to tell how they came to their own measurements.

Extension: Use either inch or centimeter cubes to measure the area of the bed. Discuss the area of the bed and the perimeter of the bed.

VOLUME MEASUREMENT

Volume is a more sophisticated concept than either length or width. Children usually gain an understanding of this later. Young children spend a lot of time putting objects in boxes, taking objects out of boxes, etc. Activities that introduce volume can build on these early childhood experiences.

Comparison

Materials needed: Small boxes of different sizes, small containers for water, tubes of varying sizes

Vocabulary: Most, least,

Directions: Give the students several small boxes or tubes or water containers) and have then predict which container will hold the most. After they have recorded their predictions, then they need to fill each container. Have them pour from one container to another to determine which hold the most and which holds the least.

Non Standard Measurement

Materials needed: Same as for comparison, plus cups, glasses, or spoons

Directions: The students will use some unit to measure the contents of each container. It might be spoons, or cups or any type of a measuring unit that is available. Have them record what each container held.

Standard Measurement

Materials needed: Same as for non standard, plus measuring cups or spoons for liquid and some media to use for the tubes or boxes.

Directions: The students will measure the number of cups or spoons in a container.

Problem: 1. Give students a bucket and have them see how many different ways they can fill the bucket (this can also enlist non-standard units) 2. Give the students several boxes in which either the inch cubes or the centimeter cubes will fit. (The teacher may have to make the boxes for this activity.) Have them determine how they can find the volume of the box or tubes.

Precision

Materials needed: same as for standard measurement

Directions: Have the students decide how they can make a precision measurement for the activities in 3. (This could be a good time to begin to show the meanings of the formulas they will use in later math.)

WEIGHT MEASUREMENT

Children are Interested in weight of different objects. This is a time when the metric scale and English scale can be used for weighing the children.

Comparison

Materials needed: various objects that have different weights.

Vocabulary: heavier, lighter

Directions: List at least 4 ways the students can determine which rock is heavier. Have them hold one object in each hand and determine which weighs the most. Continue until they have found the object that is the heaviest (or lightest). Once they have found the heaviest one, have them place one object on each side of the scale and determine which object Is heavier. Were the heaviest objects the same?

Non Standard

Materials needed: balance, unifix cubes, chain links (other objects that can be used to weight the materials used in Activity 1)

Directions: Have the students weigh the objects and record the weight of each object. Have them use two different media to weigh the objects.

Standard

Materials needed: Same as above, plus some type of weight

Directions: Place a weight on one side of the balance. Have the students find an object that will balance the scale. Continue with various objects. Use both ounces and grams.

Precision

Directions: Place an object on the scale and find out the exact weight of the object. Do this with several objects.

TIME MEASUREMENT

Time is a difficult concept to teach. Piaget states a child is developmentally ready to learn this concept by the age of 9. However, it is in many texts today and is also tested in the primary grades, so we must find ways of helping children understand the concept and gradually understand the process.

Comparison

Purpose: To help children develop an awareness of the concept of time.

Materials: pictures of different times of day - morning, afternoon, evening.

Homework Assignment: Ask parents to help children find pictures of the different times of the day.

Procedures: Students will sort and classify the pictures according to the time of day. Students must verbalize why the picture is at a particular time of day. After all of the pictures have been classified, make a large collage of Morning, Afternoon and Evening pictures to display in the classroom.

Non-Standard Measurement

Materials: Child's body, things in nature, playground equipment, etc.

Procedures: The students are now aware of the different times of the day, but are probably unaware of what happens on the playground during the various times of day. Set aside two different times during the day when students will go out on the playground or around the school and measure the shadows. Since this is a nonstandard measurement, use string or someone's foot or stride, etc. to measure the distance of the shadow in the morning. Use the same measurement instrument in the afternoon. You might want to change morning and afternoon times and see if you can find a trend for the length of the shadow. If you have a cloudy day, continue the measurement

If there is no measurement, then ask the question why?

Standard

Materials: same as in comparison, plus meter sticks or yard sticks

Procedures: Follow the same procedures as in step 2, only this time use meter sticks or yard sticks. (Older students can use the trundle wheel.)

Record the measurement in approximate yards or meters.

Precision

Procedures: Repeat Standard Step, only this time record the exact measurement of each shadow.

The above will help develop an awareness of the changes that occur during the day. As students move into telling time with clocks, there are several things that a teacher can do to help the students.

SUGGESTIONS FOR CLOCK TIME

Display a digital clock alongside of the regular analog classroom clock. Students can then compare the two times.

Cut circles and write the numbers 5, 10, 15, 20, 25, 30, 35, 40, 45, 50, 55, and 60. Attach these to the regular classroom clock. Students can then read the hour hand, and use the numbers to help them with the minutes.

Help the students become aware of the need for time during the school day. Post the times when each event during the day is going to happen. In the beginning, clocks could be posted to show where the hands should be at the time the event starts.

Note: Paper plate clock and homemade clocks may cause difficulty in telling time. These clocks do not have precision measurements. As the minute hand moves, the hour hand makes gradual movements, and this is not the case for paper clocks. A trip to a thrift shop, or sending notes home to parents, asking for a donation of old clocks, could make the process of telling time more interesting and understandable.

Time includes the calendar as well as the clock. The students can become involved with making the calendar. The calendar can also be used for place value. For each day, a stick is placed in the ones container on the right side of the calendar. When ten days have passed the sticks are bundles and moved to the left-hand side of the calendar. This can continue throughout the whole school year. Beans can replace the sticks, and when 10 beans are in the cup, glue them on a tongue depressor.

MONEY MEASUREMENT

Children receive allowances and spend money at early ages. They need to know the value of each coin and bill in order to use money. Instruction usually begins in grade one acquainting the children with the value of various coins. Even though they work with money, they do not grasp the value of the coins.

Money is sometimes very confusing. A dime is more than a nickel, but yet the nickel is larger. There is no physical relationship between coins. Children need to embark on several activities to help them understand the value of the coins and bills.

One can build the concept of money by using the "Math Their Way' concept of counting the number of days. Only pennies, dimes, and dollars were used. A penny was used for each day, and ten pennies exchanged for a dime, ten dimes were exchanged for a dollar. The dollar would represent the hundredth day of school. This could also be used with older students using the nickels and quarters.

Obtain menus from different restaurants (or make up your own). Have the student figure out what they want to eat, and then total up the bill. They must then pay that amount to a person indicated as the cashier.

Have students bring in the advertisements from the newspapers (like the department store sales bill). Give them "x" amount of money to spend to buy clothes. At this time they will have to be able to add the tax to the purchase. They must make a list of the clothes they want to buy, as well as the amount of money. This sometimes is a surprise to the students.

Students can also be introduced to the use of banking and checking with an activity such as the above. A good activity around Christmas time is to give each student $500 to buy presents for their family. They must figure out what they want to buy, write a check for each purchase, deduct that check from the checkbook, and continue. They cannot have a negative balance in the checkbook.

Another activity that is good around the holidays is to have the students list all of the things they want for Christmas, and then find the prices for those items in the newspapers or catalogs. They then must total the amount all the items would cost.

Summary of activities

All of the activities have followed a developmental sequence. This sequence enables the student to learn the meaning of measurement rather than the process of how you get the correct answer. Once students have experienced the various measurements concepts, they should be able to problem solve using different measurement.

Primary grades will probably only use the first three steps while older students will use all four steps.

Other activities that could be used to help students understand Measurement:

LINEAR MEASUREMENT

This activity involves toy cars. Students are asked to bring their favorite toy car, even the girls, and determine which car travels the longest distance. The activity can be varied by using different mediums such as sidewalk, tile floor or carpet; and also by the use of various sized ramps for giving the cars added distance.

Materials needed: toy cars, boards to use as ramps

Procedures: First have each student show his/her car and see if they can determine which ones will go the farthest. (You might want to place some specifications on the types of toy cards such as only "hot wheels' can be used.) Always start out on the flat surface, and then gradually add the ramp. The ramp can be raised gradually to determine if the height of the ramp affects the distance of the car.

Comparison

Each student will begin at the starting point and make their car go. An eye measurement will tell which car went the farthest.

Non Standard Body:

Determine how many feet the car went

The answers may vary here due to the different sized feet. You might want to choose an official foot measurer.

Non Standard Non-Body:

Use a piece of string to measure from the starting point to where the car stopped. After all the cars have been measured, place the string on a wall according to the distance and label each with the students' name. Ibis graph will help them determine the distances.

Standard

Use a tape measure marked off the inches or centimeters. A measuring stick could also be used. For instance they could say the car traveled 3 measuring sticks.

Precision (only for older students)

Use a precise measuring tool to determine the exact distance.

Extension: Do comparisons using different types of surfaces and different heights of ramps. This could also be integrated with incline plane in science.

MASS MEASUREMENT

Comparison

Materials: clay, weights, materials to be weighed, scale

Procedures: Give each student a piece of clay and have them form two equal balls that are approximately the same size. Hold one piece of clay in each hand and see if it weighs the same.

Non Standard

Place the objects on the scale and use a non-standard measurement such as paper clips, unifix cubes, washers, etc. to determine the weight.

Standard

Replace the non-standard measurements with a standard unit such as a gram or an ounce. Record the measurement.

Precision

Obtain a scale with precise measurements, and weigh each of the objects.

Question: Did both pieces of clay weigh the same?

PAPER AIRPLANES

Paper airplanes can be fun for both boys and girls. This can be an individual or group activity. Each individual or group will design his or her own paper airplane. This may take several days for them to get a model they want to fly. After each has designed his/her airplanes, s/he can follow the various stages of measurement to determine which plane flew the greatest distance.

Comparison

Determine a starting point for all airplanes. Either have each group throw their airplanes, or have one person be the official airplane flyer. Throw each plane, and look at the distance to determine which flew the greatest distance.

Non Standard

Body: Choose an official pacer, and have that person pace off the distance.

Non-body; Use string or other measuring devise to determine the distance.

Standard

Use either a yardstick, meter stick, tape measure or trundle wheel to measure the number of inches or centimeters the airplane flew.

Precision

Use a precise measurement instrument to determine the exact distance the airplane flew.

Question: Does height have any effect on the distance the planes will fly?

Extension of activities:

1. Fly each airplane several times and find the average distance of all of the flights.
2. Analyze the best flying airplanes and determine what made them fly further than the rest.
3. Make several airplanes of different materials, and determine if the type of paper affected the flight of the airplane.
4. Choose a flyer for the airplanes so each airplane is thrown with the same thrust.

Waste Paper Basket Measurement

Students like to toss their scrap paper into the wastebasket from a distance, design a contest to see who has the best aim. Place the wastebasket in a given spot in the room. Mark off Intervals away from the basket. The student must stand on the intervals to try and make the basket. The student receives the number of points from that interval. They must record the number of tries and the score for each attempt at the basket. To determine their score, they must divide the points by the number of tries.

If they miss, it is counted as a try. After a designated period of time, they determine their own score and compare scores to determine the winner.

Students following the above steps in learning measurement should be able to transfer their understanding to any measurement activities that might be presented in various textbooks. Activities can be altered to make them more interesting to the students in the classroom. Students can record their activities or write in their journals regarding the activities.

MEASUREMENT IS ALL AROUND US. HELP YOUR STUDENTS DISCOVER THAT MEASUREMENT IS MORE THAN THE MARKS ON A MEASURING TOOL.

QUESTION: HOW DO YOU DISTINGUISH BETWEEN MEASUREMENT AND GEOMETRY?

Chapter 9
PLACE VALUE

Place value is the basis for our base ten numeration system. Although we use base 10, place value can be in any base. Our place value system comes from the rule that the value of the number depends upon its placement in that number. A four can have many meanings depending on its position – it can mean a "one" or it can mean a "one hundred billion". "The Complete Graded Arithmetic" dated 1884 states

> *"The different values expressed by the significant figures, are determined by the place they occupy, and are called simple and local values. Thus, 2 standing alone or in the first place denote 2 simple units; in the second place, it denotes 2 tens as in 25; and in the third place, it denotes 2 hundreds, as in 246, etc."*

Place value must be introduced with manipulatives – not through "open your book and do "x" page." When place value is introduced by rote methods, students may become confused, and have difficulty telling the differences in value of each digit. If place value is not understood by the students, they can reverse their numbers, i.e. if the number is 47, they may reverse it and place a 74 if they do not understand that the "4" must be in the tens place and the "7" in the ones place.

An analysis of error patterns in students of various ages indicates they do not have a good understanding of place value. This happens from beginning addition and subtraction clear to operations with rational numbers. As they progress though the grades, the errors become more apparent. Our whole number system is built on the place value system.

Place value is fun, boring, confusing. These are statements of students who have been learning place value. Students who have used manipulatives find place value fun. Those who have learned it by rote or from the textbook find it boring or confusing.

Our goal in using manipulatives to teach place value is to have the students internalize the process. Careful planning, and assessment of the skills students in a particular classroom have mastered, will be the guide of where to begin instruction. Manipulatives have no special language; therefore, bilingual students can participate in lessons with the entire class. In fact, bilingual students can teach the "English only" students the names of the places in their language.

Question: *How can you use manipulatives to teach place value if you do not have any available?*

The absence of commercial made materials need not hinder the use of manipulatives in the classroom. There are many inexpensive models that can be collected and used in place of the commercial models. The type of material you will need to collect depends upon the developmental level of the students you are teaching. You need to be extremely cautious of the materials you use. Materials which make sense to you may not be appropriate for the students you are teaching. For example, money may make sense to you, and it is something that is in our daily lives. However, it is an abstract idea especially when you consider a penny is larger than a dime, and yet it is worth less.

The steps to developing the concept of place value are sequential, as is all mathematics. The first steps (especially at the lower grades) are to use materials that are easy for students to see and reflect upon. As students progress through the grades, more abstract materials will be successful. The teacher's manual, "Arithmetic by Grades" from 1897 states:

> *"Do not proceed too quickly to the abstract work because children can give answers to questions. One important purpose of using objects at this stage (Book 1) is to give a good foundation for subsequent work."*

We classify materials used to teach place value into two groups: proportional and non-proportional. Proportional means there is a size comparison; non-proportional means we assign a value to the color of the material. Always begin at the proportional level, and then move to the non-proportional.

Question: What are the levels of development in using manipulatives?

Level 1: Bundling (Proportional Model)

Manipulatives used at this level are items such as counting sticks or objects with which the students can count out a particular the number and bundle them in some manner. They can use a rubber band or plastic ties, pipe cleaners, or wires to hold the sticks together. In this bundling level, the students are able to see, for example, that they have ten sticks together in a bundle and give it the name of "ten". They also know that if they take the ten sticks apart they will have ten individual sticks.

Bundling sticks did not come as a result of the new math or the new new math. In the teachers manual "Arithmetic by Grades" on page 23 it states:

> *"In teaching of number to 100 by means of sticks . . . . The purpose of such teaching is to give the children a good foundation of knowledge of numbers and their expressions. As before every ten of sticks should be bound into a bundle and the number of such bundles in a number should be called so many tens, while the single sticks should be called units or ones."*

Substitute materials: straws, coffee stirrers, popsicle sticks, pipe cleaners, toothpicks, plastic silverware, strips of paper, or any material that can be bundled together to make a group.

This step should be used with primary children when they begin to learn about place value. Again this does not necessarily mean a base 10; it can be any base, and taught in any language.

Level 2: Numeration Blocks (Proportional Model)

This model is a proportional model, as it still has a size relationship. Although this is not a bundling model, the relationship between the hundreds, tens and ones can easily be seen. In this model, the students exchange ten ones for one ten, and ten tens for a hundred. This is the model often pictured in textbooks.

One of the first models pictured in books was in the "Complete Graded Arithmetic, Oral and Written" dated 1884 on page 11. The blocks are pictured in a checkerboard pattern. The book states:

> *"The order of units increase and decrease by the uniform scale of ten. Ten units of any order make a unit of the next higher order. Moving a figure ten times to the left increases its value ten times. Moving a figure one place to the right diminishes its value ten times."*

There are substitutes for this model. Bean sticks are popular. Students glue ten beans on a stick for a group of ten, and ten sticks together for a hundred. They use the single beans as ones.

Other models which are considered proportional models are beans and cups. Small medicine cups can be used and when students collect ten beans they place them in a cup. The beans have a rule – they can only be placed in a cup when there are ten of them. If there are less than ten, they are considered a single bean.

Beans make an inexpensive model, but, buttons, pasta, popcorn, sunflower seeds, small rocks, etc. can be used in place of the beans.

Graph paper cut into hundreds, tens and ones, can also be used. It is suggested that either the 2 cm. or the inch graph squares be used, as they are easier to handle than 1-cm squares.

Level 3: Chip Model (Non-proportional)

This is probably the easiest model to obtain, and also the most abstract. Students need to be at a higher developmental level in order to grasp the concept at this level. In this model the students must remember the color and the value of the chips, i.e., yellow may be "1"; red may be "10"; blue may be "100", and green may be a "1000" i.e. In this case, 10 yellows exchange for 1 red; 10 red for 1 blue; and 10 blue for 1 green. This concept requires a higher level of thinking.

Warning: Just because a model makes sense to the teacher does not mean it makes sense to the student. A model must fit the developmental level of the student. For example, a group of bilingual third graders, who had never used manipulatives, were introduced to the colored chip model. This model made sense to the teacher, but was too abstract for the students. A different third grade class at the same school, which had a similar population of bilingual students, used the beans and cups, wherein, they placed 10 beans in a cup and called it a group of ten. They understood a ten was another name for ten ones. They actually placed the ten beans in the cup and moved them to the ten column.

<u>Substitute models</u>: colored squares of paper, pom-poms, colored lima beans, colored macaroni, colored paper clips, etc.

As students move beyond late third grade or early fourth grade, it is easier to represent large models using the chip method. You are only limited to the number you model by the number of colors you have available.

Developmentally, students in the primary grades should use the proportional models. Once they reach the concrete operational stage, the non-proportional models may make sense to them. If older students are experiencing difficulty with the nonproportional model, use the proportional model to help them develop an understanding, and then move back to the nonproportional model.

Give the students time to develop the concept of place value, whether at the proportional level or the nonproportional level. They need a variety of hands-on experiences in reading, writing and modeling numbers.

Experiences in the Pre K – 2 Grade Levels

The NCTM proposed Standards states that during the pre K to 2nd grade span, mathematical growth is remarkable. The teachers of this age student, whether school teachers or parents, need to be aware of the importance of the earliest beginnings of mathematical reasoning. They need to realize that developing a solid foundation at this level nature's mathematical learning, and sets the stage for later development. Teachers of this age group need to understand child development, need to know how to build a strong conceptual framework, and how to build on the child's past experiences. They also need to be able to nurture the student's natural inclination to solve problems, and present settings, which allow for free exploration.

ACTIVITIES FOR PRIMARY STUDENTS

Students need to have a variety of experiences prior to doing operations with multidigit numbers. The teacher needs to carefully plan activities that will fit the needs of the students in the class. S/he needs to assess prior learning; consider the languages spoken at home; and determine what s/he wants the students to learn. Once the teacher has carefully analyzed the students, activities can be planned. Some of the activities which have proven to be successful with primary aged students could be consider games, rather than learning activities.

Activity 1: Banker's Game *(prerequisite to addition with regrouping)*

Materials needed: *counting sticks, numeration blocks or chips; place value mat (made from 9 x 12 piece of paper folded in half lengthwise with a line drawn down the fold – the word tens is placed on the left side of the paper and the word ones is placed on the right side of the paper.) die*

Procedures: *Students will roll one die (random number generator) and collect that number of ones, i.e. if a 5 is rolled, the student will collect 5 ones. Once ten ones have been collected, they are exchanged for one ten. If sticks are used, they will bundle them into a group of ten once they have ten ones. Play continues until they have collected a predetermined number of tens (e.g., six tens). The winners are the ones who can tell the number of tens and ones they have. Students will probably play the game several times after they have completed their work, or during activity time.*

Variation: *After the students have played the game several times and understand the game, they can begin to add numbers to their actions. This will help them "bridge the gap" between the abstract and the concrete.*

Activity 2: Gambler's Game *(prerequisite to subtraction with regrouping)*

Materials needed: *Same as Activity 1*

Procedures: *Students begin with a certain number of tens, e.g., 6 tens. They will roll the die and take away the number they rolled. The first thing they must do is to exchange a ten for ten ones. Students take turns rolling the die, and taking away that number of ones. The goal is to reach 0.*

Bridging the Gap: *The next step is to help the students bridge the gap between the concrete and abstract. Students need to use numbers to represent their actions. Two sets of digit cards (0-9) can be used. One for the ones and the other for the tens. Once they make an exchange, they will change the numbers to match the manipulatives.*

Activity 3: Tens and Ones

Materials: *manipulatives, place value mat*

Procedures: *The teacher places a tens and ones chart on the overhead. S/he proceeds to write a number in the tens column and the ones column. Students are asked to model the number. Each student should be able to model the number on his/her chart. As the students are modeling the number, one student is chosen to go to the overhead and model the number. This gives the rest of the class a chance to check their own work. The main emphasis is placed on modeling the number.*

The child who has modeled the number then writes a number on the overhead and chooses another child to model the number. This allows the children to be the teacher and the teacher does a walk-about assessment to determine which children are successful and which need help. Remediation is taken care of before it becomes a habit.

Activity 4: Write the Number

Materials needed*: magic slates, individual chalk boards, number charts and digit cards, zip lock bags filled with finger paint, or some soft material on which students can write.*

Procedures: *The teacher places some tens and ones on the chart. Students look at the manipulatives and write the number for the model. This allows the teacher to do a walk about assessment.*

The same procedure can be used as in activity 3.

Activity 5: Listen to the Number

Materials needed*: manipulatives, recording materials, digit cards (cards with the numerals 0-9 on them)*

Procedures*: The teacher asks the students to model a given number (e.g., 45). The students place 4 tens and 5 ones on their desk. They then place the numbers 4 and 5 to represent the tens and ones. One student is asked to model the number on the overhead to allow the class to check their answer. As the student displays the number on the overhead, the teacher does a walk-about assessment.*

The teacher should mix the numbers given. Some numbers should be expressed as tens and ones and others as ones and tens, e.g., "I have seven ones and 8 tens."

LANGUAGE OF PLACE VALUE

Once the children understand the above activities, they can begin to write stories, poems, or riddles about various numbers. Van De Walle (1990) suggests students make up riddles such as "I have 23 ones and 4 tens: "Who am I?" Students will model the number and make the necessary changes to find the new number.

Writing Numbers

The primary grades have always emphasized writing the numbers to 100 or to 1000 or to 10,000. It was assumed that if the students know how to write the numbers, they understand the numbers. However, this is not always true. Mathematics Their Way suggests an activity that will help bring meaning to this activity.

Math Their Way Activity

Materials needed: beans, unifix cubes, or any individual material that can be used for grouping mats (left side is plain while the right side has 10 boxes drawn on it, flip cards

Tens	1		0	Ones
			☐ ☐ ☐ ☐ ☐	
			☐ ☐ ☐ ☐ ☐	

Procedures: Students begin by placing an object on one of the boxes on the right side. After a student places an object on the box, they flip a card and match that number with the same number of objects on the chart. Once all of the boxes are filled, the student places a group of ten on the left-hand side. The flip card now reads 1 ten and 0 ones. They continue for a period of time.

Extension: Use a roll of paper such as adding machine paper to record the numbers. They first write a zero on the paper. They add one object and write a "1". They continue until they have ten. When they have ten, they place it in a cup and continue with adding ones. Once they have ten cups with ten objects in each cup, they will place it in a larger paper cup to represent the number 100.

Activity 6: Game: Whole Class Bingo

Materials needed: plain piece of paper, pencil, and markers

Procedures: Have the student fold their paper into 16 squares (or any number you decide is appropriate). Place 16 numbers on the board. The students place the numbers in their squares. The teacher calls out the numbers such as 8 tens and 3 ones or 3 ones and 8 tens. The students place their markers on the appropriate square. The teacher can also use riddles for the number, forcing the student to use their listening skills to decode the answer.

Activity 7: Numbers Have Different Names

Materials needed: manipulatives, expanded cards (cards with 2 digit numbers; 10 – 90 on each one; cards with 1 digit numbers 0-9 on each)

Example: 24 = |20| |4|

Procedures: The teacher shows a 2-digit number and asks the students to choose the two cards that would make the number, i.e., if the teacher shows the 2 tens and 4 ones, a student would show the 20 card and the 4. The student will read the expanded notation of the cards – 20 + 4 = 24. The student then places the 4 on top of the zero to make the compact number 24.

> Once the students understand expanded notation, ask them if there are other ways to say 24, e.g., 1 ten and 14 ones, or 24 ones. Once they have mastered 2-digit numbers, move to three-digit numbers.

Extension: Make a set of concentration cards to be placed in a learning center. Students will play a concentration game and at the same time gain experience with numbers, which will be necessary when they begin to do various operations.

Game: A game could be created in which students must find their partners, i.e. one student will have a card with 24 on it, and the other will have a card with 1 ten and 14 ones. Once they find their partner, they must tell how the numbers are alike and how they are different.

These cards could be placed on the student's back, and they would have to ask questions to discover their number. They can only ask questions which can be answered with a yes or no!

EXPERIENCES IN GRADES 3-8

Teachers have a major goal at this level – make it interesting and meaningful for the developmental level of the students. They need to help these students understand that they are responsible for their own learning. They need to be able to examine problems, ask questions, and consider different strategies. The goal at this level is to make sense of mathematical ideas and fit them together.

The material presented in grades 3-5 builds on what has been taught in the Pre-K to 2nd grade program. Students at this level are becoming more sophisticated in their reasoning ability and need to be challenged but yet supported in their learning.

The goal of mathematics instruction in the middle grades is to lay a solid foundation for secondary school and beyond. The focus should be on continuing to build on this strong foundation in order to use mathematics to deal with quantitative matters throughout their lives. It is the job of the teacher to be aware of the developmental level of their students and to provide an environment that is conducive to learning and one where students form positive opinions regarding themselves as learners of mathematics.

Activities for Upper Grades

Students in the 4th - 6th grade should already have a basic knowledge of place value. Even thought they have had many experiences with numeration, you cannot assume they understand the concept. Teachers should do an assessment of the student's prior knowledge, and design instruction to meet their needs. At this point, a game similar to the Banker's game or the Gambler's game could be used.

Activity 1: Bankers Game

Materials needed: *manipulatives die (dice)*

Procedures: *The first time they play the game, have a review game. They will roll a 6-sided die and collect that number of ones, and once they have ten ones, exchange it for a ten. This will help them review the game, and also give the first-time players a chance to get acquainted with the game.*

From this time on, you will build larger numbers - a 20-sided die will make larger numbers, or take 2 six-sided dice and have them multiply the numbers together. The numbers on the dice then will depend on how large of a number they are building. For example, if you are building numbers to 300, you would want to use the 20-sided die. If a 17 is rolled, they will collect 1 ten and 7 ones. This will also encourage the use of mental math to avoid time spent in exchanging. If numeration blocks are available, they should probably be used for the first few activities. After that, change to the chips.

Activity 2: Gambler's Game

Materials needed: Same as for the Banker's Game

Procedures: Have the students take a certain number of 100 blocks, e.g., 3. have them take away what they roll. After they have reviewed this game, let them change to the chip model.

Activity 3: Digit

Materials: digit cards (half of a 3 x 5 card), digit mats (3 x 9 paper divided into 3 parts)

Digit Cards 2 x 2 cards with one digit on each card (0, 1, 2, 3, 4, 5, 6, 7, 8, 9]

Hundred	Tens	Ones

Digit Mat

Procedures: *Each student makes his/her own digit card and digit mat. Once they have made their cards, they shuffle them very well, and place them face down on their desk. The game can be changed to make the highest number, lowest number, number closest to 500, etc. In building any number, they will turn over their first card and decide where to place it. <u>ONCE IT IS PLACED IT CANNOT BE MOVED</u>. Play continues until all three cards have been played. Students then take turns reading their numbers.*

Variations: A list of the numbers can be written on the board, and the students asked to order them from largest to smallest. They can also determine the difference between the largest number and the smallest number, or find the average of the number, etc.

Activity 4: Expanded Notation with Cards

Materials needed: a 4-section card, 3-section card, 2-section card and 1-section card, digit cards, cards with "+" sign on them

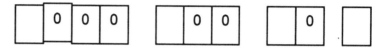

Procedures: Students will place the mats in front of them. If they are building a 4-digit number, they will place a number on the first space of each card, and place the "+" signs between each of the cards. They will then read their number sentence, i.e. 4000 + 200 + 30 + 4 - they will then stack the cards on top of each other so they have the compact number 4234. They will then read the compact number.

Variations: They read the cards as 42 hundreds or 423 tens or 4234 ones. Ask them to continue building numbers and reading them, both in expanded form and in compact form.

The teacher gives the students a number, and they place it on the expanded cards.

Roll 4 dice and build a four-digit number to match the numbers on the dice. Write a sentence about the numbers drawn, and write the compact number. Roll the 4 dice and make the highest number, or the lowest number.

ACTIVITIES BEYOND THREE DIGITS

Students in the upper elementary grades should have a solid background in place value. However, since we can never assume anything, we will need to make a quick revisit of the basic skills. We need to assess their knowledge, and then plan instruction to meet the assessment needs.

Many students have difficulty reading large numbers. They do not know where to start, which numbers to read together, or how to say those numbers. They also are unsure of what each digit stands for. When they are first making larger numbers, color squares of papers can be used to model the numbers. The chip-trading model should be followed.

Every three numbers is called a "period". Each period contains hundreds, tens, and ones. These are repeated for each period that is added to the grade level. We usually go up to trillion in reading large numbers (since this is related to our national debt). The 1884 textbook stated the periods were:

> "trillions, quadrillions, quintillions, sextillions, septillions, octillions, nonillions, decillions, etc." It also stated to express numbers as figures "start on the left and write the figures in a given order in

succession towards the right. If any orders are omitted, supply their places by ciphers and separation tenths from units by a decimal point."

Note: they are using the decimal to separate periods, instead of the comma. There are some countries that still use this system. If you have students from other countries, check to see if they use the comma or the decimal prior to counting it wrong.

Following this suggestion, we can help students be successful in reading the large numbers by helping them build mats for each period.

Reading large Numbers

Materials needed: 2 sets of digit cards, cards (folded to make them look like tents) with the words with "hundred(s)", "thousand(s)", "million(s)", "billion(s)", and "trillion(s)" and 4 commas cut from rectangles (students must cut their own commas).

Procedures: Begin with one period card, and have the students replay the digit game. After they have played this several times, add a second period card. Place the word "hundred(s)" over the right period card, and the word "thousand(s)" over the left-hand period card. Place a comma between the two period cards.

Thousand(s)				Hundred(s)		
Hundreds	Tens	Ones	,	Hundreds	Tens	Ones

Students place their digit cards on the six spaces. The same rule is followed. Once the cards are placed, they may not be moved. The student reads the numbers

- Start on the left
- Read the first three numbers
- When you come to a comma, read the word above (in this case, when you come to the comma, you will read the word thousand)
- Read the next three numbers
- Since there is no comma after the hundred(s) period card, the word hundreds is not read

The students can build up to a 15-digit number, each time separating the periods by a comma, and reading the word above for each period.

Homework: This would be a good homework assignment. The students can take their cards home and practice making and reading large numbers with their parents. If the students are bilingual, and a language other than English is spoken in the home, they can add the words in their language.

Activity 1: Comparison of large numbers

Materials needed: *same material as in Activity 1, signs <, >, =*

Procedures: *students build two multidigit numbers and place the correct sign between the numbers. Students then record their two numbers.*

Variation: *This would be a good homework assignment. Students could use the same procedures as in Activity 1, only record the numbers compare then and use the correct sign.*

Activity 3: Rounding numbers

Materials needed: *digit cards, charts, period names, colored squares, and marker*

Procedures: *Students will build a multi-digit number; model the number using colored squares of paper. They will then place their marker (small clothes pin) on the tens (or number they want to round to) place to indicate they are rounding the number to the nearest ten (or number they are rounding to).*
Rule: *<u>If the number is five or larger, round to the next highest number. If the number is below five, then do not change the number.</u> Students will then place zeroes on the numbers following the one that has been rounded.*

It is best to start with 3 digit numbers and then slowly add more digits.

Worksheets: *A pair of students will build numbers using their digit cards. They will compare the numbers; record their numbers on a piece of paper. They will then decide which number they will round to and place a marker on that number. They will follow the rounding rule, and make a new number. They will tell which number they are rounding to, and then write the new number. A die or card could be used to tell which number is to be rounded. Once they have had experiences similar to these, then they can do the pages in the textbook.*

Example: Use a 5-digit number and round it to the nearest ten thousand. The number is 68,321. A clothespin is placed on 6 to indicate it is the number to round to. A check of the number to the right indicates it is larger than 5. The 6 is then changed to a 7. The numbers 8321 remain, however, since we are rounding, we will need to place zeroes on these numbers. Our rounded number now is 70,000.

Numbers Less Than One (Decimals)

Decimals are numbers, which extend place values to less than one. A dot or period usually separates the whole number and the decimal. The numbers to the left are whole numbers and the numbers to the right are called Decimals. The dot or period is always read as "AND". The name decimal comes from the Latin work *decem,* which means ten.

The first number to the right of the decimal point is called tenths. The second number is called hundredths, the third number thousandths, etc. An exercise from the 1884 book gave the following example of how to read numbers with a decimal.

651,780,900,240,785.324

Six hundred fifty-one trillions, seven hundred eighty billions, nine hundred millions, two hundred forty thousand, seven hundred eight five AND three hundred twenty four thousandths.

Question: How is this different from what we teach today?

Decimals May be Different in other Countries

Assessment of the student's assessment of decimals will help teachers determine how to teach decimals. In some countries a comma is used in place of a decimal point. In other countries the decimal point is placed in a different location, and in looks like an exponential. If all of your students are from American schools, they probably have the same background. If the students have had any schooling in another country, they should be given the opportunity to share how they express decimal numbers in their country.

The first account of the decimal system in print was 1585 - so decimals are the newest concept introduced into mathematics. They came about when there was a need to express numbers less than one but greater than zero. They allowed us to extend our numbers system to the right and left of the "ones" place.

Students seem to have more difficulty with decimals than other numbers. Part of the reason for this is they were usually introduced through rote memorization. Once students reach this point, it is assumed they do not need hands-on experience; however, this is not true. As with whole numbers, time needs to be spent comparing decimal, finding other names for decimals, and making decimals to help the students gain the concept. Decimals and fractions are related. Work with fractions can help to bridge the gap between the two concepts.

Activity 1: Comparing Decimals

Materials: *Numeration blocks (hundreds, tens, and ones)*

Procedures: Students may have experienced the use of numeration blocks in the primary grades, but now they are going to use the hundreds block as the unit, the ten's block as .1 and the one's block as .01.

Students will place the .1 blocks onto the one and see how many it takes to cover the block. They will them place the .01 on the block and see how many it will take to cover the blocks.

Students roll a die with the numbers .1, .05, .01, .17, .24, and .09 on it. Each student will have a unit (hundred) block. They will take turns rolling the die and placing that number of units on the block. The goal is to cover the entire block.

Students will have decimal mats, one divided into ten equal parts, and one in 100 equal parts. The students will color in half of each, and write a sentence to compare the two mats.

Activity 2: Decimals

Materials: numeration block model, 3 dice – one with a decimal point before the numbers, and two with the numbers 0-9 placed randomly on each (6 numbers of each die)

Procedures: Students will roll the three dice and build a 3-digit number. If the numbers rolled are 3, 7, and .0 they can either build a model for .037 or .073. The goal of the game is to build a cube, so they would take .073. If the next roll was a .063, they would add 6 longs and 3 units. They now have 13 longs and six units. They can exchange 10 of the longs for a unit. Play continues until someone has a cube.

Activity 3: Digit less than one

Materials: digit cards, words tenths, hundredths, thousandths, decimal point, digit mats

***Procedure:* Place the decimal point on the digit mat. Decide if you want a small digit number or large digit number.** Students play the game the same of the regular digit game. They practice reading their digit numbers.

Remember: Place value is the basis for all of our mathematics. Teach it well and teach it thoroughly!!

Chapter 10
BASIC FACTS

Basic facts have long been seen as the mathematics program. All attention has focused on helping children learn their facts and mastery was judged by being able to pass timed tests. This is a myth that still exists today.

Math reform has moved from math as a total subject to emphasis on basic facts and timed test to developing an understanding of the concept so students will be able to problem solve. Even the way basic facts are taught is beginning to change. During the last state reform, we moved away from a sequence of memorization and excessive drill and practice to a sequence of understanding and helping children discover the facts for themselves. The math guidelines have moved to memorization, and have moved concepts that were once third grade concepts to the first grade. We are asking more of our students sooner and are not concerned with understanding.

What are basic facts?

Basic facts are operations on single digit numbers (0 + 0 to 9 + 9 in addition or 0 x 0 to 9 x 9 in multiplication). When we begin to use numbers beyond 10, we are using multidigit operations. This is the beginning of the number system that combines one or more digits.

Rathmell (1978) in the 1978 NCTM yearbook stated there were three components to teaching the basic facts. (These statements still hold true for the new century.) These components are:

1. Concrete Materials

Concrete materials should be used during initial instruction no matter the grade level. The main purpose of the concrete materials is to develop the concept for the operation, with the appropriate language and symbolism. Concrete materials become a referent for work involving operations - they provide the link to connect real world to problem-solving situations. These materials are also used as the "proof" of a problem.

Concrete materials enables students to talk about what they are doing and what they see happening. This stage helps build a referent for each symbol.

2. Role of Thinking Strategies

Teaching thinking strategies helps children learn mature strategies that are useful and effective to solve harder facts. It enables them to discard the concrete materials. Thinking strategies are important for helping children recall facts. They facilitate organization and make whole sets of facts more coherent.

3. Role of drill

Drill is an essential component of instruction. Practice is necessary to develop immediate recall. Drill on hard facts should be delayed until students have thinking strategies.

Thinking strategies help children derive answers to fact problems.

ADDITION AND SUBTRACTION OF WHOLE NUMBERS

What does a student need to know in order to add and subtract?

1. Attributes
- They must recognize the specific characteristics of a set, as well as be able to distinguish between the various signs.

2. Seriation (Ordering)
- Students who seriate and classify will be exploring likeness and differences.
- Students must be able to place objects in order according to given criteria (length, width, height, weight, diameter, etc.) in order to abstract number and words. Seriation is necessary to recognize the difference between numbers and letters.

3. Relationship Between Two Sets
- Matching and comparing sets
- Noticing similarities and differences is an important learning process. By close observation and measurement, and by physical comparison, the student's perception is developed.
- 1 - 1 correspondence between sets
- 1 -many correspondence (< > =)

4. Ordering sets

5. Counting
- Counting from I - 1 –2 –3 - 4- 5 - 6 - 7
- Counting on from a given number - 6 - 7 - 8 - 9 - 10
- Counting back from a given number - 9 - 8 - 7 - 6
- 1-1 Principle (each item in a set to be counted must be assigned one and only one label.)
- Cardinal principal - Final word assigned to a set identifies the numerosity

6. Joining and Separating sets.

What is the developmental level of students learning to add and subtract?

Most children are still in the preoperational stage, as listed in the readiness section. Some are beginning to move toward the concrete operational stage.

What are the characteristics of the concrete operational stage?

- begin to think logically
- reverse thought process
- mind is like a moving camera (they can take 2 objects and 3 objects and place them together to get five objects at the same time they can still visualize the 2 and the 3
- order objects according to attributes
- begin to understand the operation
- can conserve quantity

X X X X X X

X X X X X X

The student is asked if the two sets are the same, or if one set has more or less than the other set. The student will usually say they are the same.

The sets are then rearranged:

X X X X X X

X X X X X X

The student is again asked if the sets are the same, or if one has more or less than the other.

<u>**Non-conservers**</u> - say more in top row

<u>**Conservers**</u> - there are the same, you just spread them apart

Non-conservers will have a difficult time memorizing basic facts, while conservers will not have that difficulty.

Conservers notice reversibility

    Notice 2 + 3 and 3 + 2 give the same answer.

This eliminates approximately 50% of the facts that need to be memorized.

What skills do students need to start be successful in learning the basic facts?

Students practice in working with attributes, patterns, seriation, ordering sets, classifying sets, and counting have been prerequisite skills for basic facts. Therefore, the next step is to look at the relationship between sets.

<u>*Don't forget the period of free play before using any manipulative for instruction.*</u>

Some activities that help build an understanding of addition and subtraction - is a process called **equalization.** Although most addition and subtraction is introduced through joining and separating of sets, equalization is a process which allows the student to decide whether they want to add or subtract. Rand McNally, in their series called Developmental Mathematics Processes; (DMP) introduced this process.

Although most textbooks do not use this process, it is a readiness step that should be completed prior to introducing textbook pages. Informal research has shown that by using these methods students understood the concept, and the textbook pages that were used for practice proved to be easy and were completed with speed and accuracy.

SET COMPARISON

Equal/Not Equal

Set comparison is a prerequisite step to addition and subtraction. Set comparison is a preschool activity which students have engaged in prior to coming to school. We need to build on this experience and expand it in the primary grades.

Geometric solids allow students to gain the concept of addition and subtraction without having to assign a number to the operation. This step is important, as we want students to understand the concept prior to assigning numbers.

Activity 1: Comparing sets

Free Play: Place balances and geometric solid pieces in learning center one week prior to instruction to enable the students a period of free exploration where they will begin to abstract the characteristics of the manipulatives.

Preactivity: Tell a story about a character that does not like anything equal. A character called "Grumpleflurg" was used in the DMP series. Grumpleflurg was a weird looking character who had nothing symmetrical or equal. His face was not equal, his clothing was not equal, and his feet were not equal. In fact Grumpleflurg would have a temper tantrum when he found anything that was equal. (Although I have mentioned Grumpleflurg, a character could be made up that would accomplish the same thing as Grumpleflurg A picture could be made of the character and shown to the students while telling the story.)

Vocabulary: equal, not equal, balance,

Materials needed: balance, geometric solids with letters on

Direct Instruction: The teacher asks the student to place two objects on the balance. Each group will have different objects on the balance; therefore, each balance will be different. The teacher asks the different groups if one

> *object is heavier than the other. She then asks the students whether Grumpleflurg would be happy or have a temper tantrum. This is repeated several times before the teacher asks the student to place certain geometric pieces on the balance - such as a piece marked **A** and a piece marked **B**. A reference would be made to Grumpleflurg and students would be shown a sign to enable them to record their experiences. The* equal sign (=) and not equal sign [≠) would be introduced.

As the students interact with the objects, they begin to abstract the idea of equal and not equal. Students who are physically involved in their learning will acquire certain concepts and those concepts will have personal meaning to them. Through the various activities the student will also begin to build visual images.

Paper and Pencil Activities:

  <u>Materials</u>: Balances, geometric solids, and worksheets

  <u>**Directions**</u>: The student completes a worksheet, which is designed, something like this:

The student places the geometric solids labeled "A" on one side of the balance and the solid labeled "B" on the other side. The students then writes an = or ≠ sign between the two letter. The student then proceeds to the next box and completes the action.

<u>**Extension Activities**</u> Instead of using the geometric solids use other objects that are available in the classroom or other things which the teacher provides, i.e. scissors, erasures, thread spools, blocks, etc.

All of the above activities do not require numbers. The following activities will be the first they use that require knowledge of numbers.

Activity 2: Comparing Sets

<u>**Materials needed**</u>: *unifix cubes, chain links, or other materials that can be used with balances.*

<u>**Direct Instruction**</u>: *Give the students directions similar to the following:*

Place 5 cubes on one side of the balance, and 5 chain links on the other side. What sign would you use? Would Grumpleflurg be happy?

Do several similar activities with the same manipulatives. Although they can apply numbers to the unifix cubes and the chain links, they are not the same; therefore, they will have to decide how to equalize the balance - without numbers.

> *After they have had many experiences with the cubes and chain links, move to chain links only, or unifix cubes only. Place 7 cubes on one side of the balance, and 3 on the other. etc. The children can now begin to write how they made their balance equal. Once they understand the concept, they are ready to complete some paper and pencil activities.*
>
>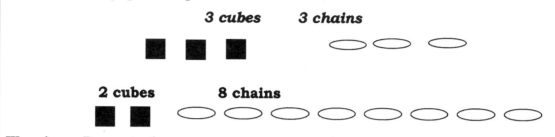
>
> **Warning:** Do not do premature paper and pencil activities. Drill before experience could prove dangerous.

Paper and Pencil Activities: A worksheet similar to the following can be used:

| 5 | 3 |

Greater Than Less Than

Once students understand the symbols of equal and not equal, they are ready for the next step: greater than, less than. Many characters have been used to introduce this concept, such as the alligator and Pac man. The character that will be referred to in this section is Greedy Duck (DMP character).

- **Preactivity** Tell the story of the character you will be using. Greedy Duck is always greedy and wants the most of anything. If he were *given* the choice of two pots of honey - one small and one large - which would he choose? (Objects can be shown and the students asked which Greedy would choose.) One day Greedy went to visit Auntie Duck. Auntie Duck made chocolate Pancakes for them - one small pancake and one large. She asked Greedy which one he *preferred,* and he naturally said *"The bigger one'*

Students should have a variety of experiences working with the concept before they are held responsible for the terminology. The teacher will be using the terminology as she provides direct instruction to the students. If we force the terminology, students may become confused. Once they are secure with the concept, the terminology becomes automatic.

Vocabulary: Greater than, Less than

Materials needed: Balance, objects to use on the balance similar to those used for equal, not equal.

- **Directed Instruction**: Use the same activities as used for equal and not equal, Introduce the signs "< >" and refer to Greedy's mouth, which is always open, to the larger object. Classroom activities such as showing objects in the classroom *and* having a student use the sign to tell which Greedy eats.

This can also be extended to have students In the class stand in the front of the room and tell them Greedy wants to eat the tallest one -which one would Greedy eat?

Paper and Pencil Activities: The same worksheets can be used for" equal – not equal", except for changing the signs to < >.

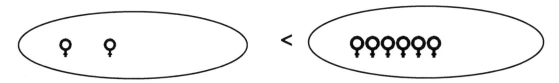

ADDITION OR SUBTRACTION

Which is your preference?

Once students have completed a variety of activities similar to the ones listed above, they are ready to begin activities which focus on addition and subtraction. The first activities should use the geometric solids, to reinforce the fact that something can be added on or taken away from the solids - thus focusing on the concept of equalization, rather than on a given quantity.

If you have children whose first language is other than English, you might want to use only one term, such as add or subtract, and not use plus or take away until they are comfortable with the terminology. You can also make a chart, which shows the word in their native language and the word in English. The main complaint we have from students, who have had some of their schooling in another country that the United States constantly changes the names they use for the concept of addition or subtraction.

Activity 1: Equalization

Vocabulary: *some (which will later become the missing addend), equalization, add, take away*

Materials needed: *balances, unifix cubes, chain links, washers, etc.*

Writing in Math: *The students will make up their own addition and subtraction sentence.*

Example: *Place 3 cubes on one side of the balance and 6 cubes on the other side. Ask the students to equalize the balance. They will either add to the 3 or take away from the 6. Ask the students to write a sentence to telling about their actions – i.e., "I added three to the three" or "I took 3 away from the 6". (An algorithm is not required at this time - only the students' written language.)*

Once the students have had the opportunity to write many sentences by themselves, they are ready for direct instruction. Following direct Instruction. the textbook pages can be incorporated.

Direct Instruction: *Ask the students to place 6 cubes on one side of the balance and place 3 cubes on the other side of the balance.*

Several problems are completed in this manner.

This can also be considered the beginning of algebra activities.

Activity 2: Number sentences

(When a student begins to write their own equations, it is referred to as a number sentence.)

Materials: *balances, unifix cubes, chain links. or other materials that will balance on a scale (the objects should be the same on each side of the scale), die.*

Directed Instruction: *The student will role a die and place that number of objects on one side of the balance. They will role the die a second time and place that number on the other side of the balance. They will then equalize the balance and write a sentence about their actions.*

Activity 3: Number sentences using Cuisenaire Rods

Question: *What colored rod will make the light green rod equal to the yellow rod? (red rod)*

Teacher records Light Green (lg) + red (r) = yellow (y)

As the students become more familiar with the rods and are using them in their work, the teachers can then record 3 + ☐ = 5 and the students will place a 2 in the box.

We use the same method to equalize a subtraction problem. The student will again model the five rod and then place the three rod on top.

Question: *"What color would you need to take away from the five (light green) to have only 3 (yellow) left?" The students know that the 3 plus something else equal five. Again they will experiment and find they need to take a 2 (red) away to have 3 left. Again this builds the equalization process.*

The teachers need to feel comfortable with these manipulatives prior to using them to teach lessons. If the teacher is insecure, then the instruction will suffer and students will not learn the concepts.

Another method that can be used with the Cuisenaire rods is called the "family of 5" etc. The student takes a 5 (Light Green), and finds the various other combinations that are the same length as a five rod, i.e. - purple + white; red + yellow, etc.

Learning Center Activity,

Materials needed: 2 dice with the ten colors (dots colored the same, as the rods pasted on a regular die will do)

Directions: Students roll the 2 dice and collect the appropriate rods. They then write an equalization sentence for the rods.

Again, worksheets similar to those used for the other operations can be used.

Painted Lima Beans

Vocabulary: some (which will later become the missing addend), equalization, add, take away

Materials: Lima beans spray painted on one side. Two circles drawn on a sheet of paper.

Directions: Place five red lima beans in one circle and three white lima beans In the other circle.

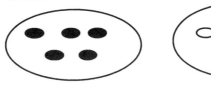

Teachers ask students how then can make the number of lima beans equal. Again following the same procedure as before, they can add or take away.

The students though, should reply that you could add two to the white, or take away 2 from the red. Their number sentences might be:

5 - ☐ = 3   OR   5 = 3 + ☐

Independent or Seatwork Activities The students will place some beans in a container - i.e. 7 lima beans - and turn them out on a piece of paper. The students then write sentences to describe the number of red beans and white:

3 + 4 = 7   **2 + 5 = 7**   **6 + 1 = 7**

Or the students can begin with the 7 and take away the red beans:

7 – 5 = 2   **7 – 4 = 3**   =   **7 - 1 = 6.**

Bridging the Gap: In all of the above activities, the teacher has modeled the writing of the actions on the board or overhead. The students need to follow the teacher's examples and begin to record their own actions.

When students write their own sentences, they can use either addition or subtraction. The sentences become their personal problems, which they can refer to as time goes on.

Building Visual Images: As students work *with* the various objects, they will begin to build their own visual images. They will gradually be able to make mental images of how a five and a three can be equalized. This is one step they must do for themselves; the teacher cannot do this.

Literature and Mathematics

Materials needed: Literature Book or story. Lima bean ladybugs.

Lady bugs and Math

Activity 1: Equalizing Ladybugs

Vocabulary: add, take away

Materials: ladybugs made from lima beans, book, and flowers

Directions: Have the children use two of the flowers that have been made (or make them at this time). The children will place the ladybugs on two flowers and write a sentence about the ladybugs or they will place some of the ladybugs on the flower and have some fly away

The children can use two dice to determine how many ladybugs to place on each flower. They can then determine how to equalize the numbers on each flower. They will write a story about their actions:

i.e. "I had five ladybugs and two flew away. Now I have 3 ladybugs."

Activity 2: Write a problem

Materials needed. paper and pencil

Directions: The children should be asked to write problems about the ladybugs. If the children cannot write the words, let them use rebus (small pictures of the object) pictures to represent the ladybugs and flowers or they could dictate a problem to someone, or use a tape recorder. The problems are then placed in a learning center for other children to solve. Note: The children will also need to make containers of ladybugs to go with the problems. The children's names are placed on the problem and they are asked to correct all papers for their problems.

WHOLE LANGUAGE AND MATHEMATICS

Although whole language is not an accepted method of teaching reading, it can be useful in teaching mathematics. Whole language has several important components: talk, literature, writing, and drama. These important components are also important in mathematics. Literature is an important component in introducing a concept, particularly at the primary level. In the following activities, a story has been used to introduce the concept, and activities will follow that integrate the important components of whole language (mathematics) teaching.

Activity 1: Add Those Frogs

Materials needed: origami frogs, lily pads, and dice or digit cards

Directions: Roll the dice (or turn over a digit card) and place that number of frogs on a lily pad. Repeat the activity. Write a number sentence for adding the frogs.

Example: If a 4 is rolled, place four frogs on the lily pad. If a 5 is rolled the second time, they add 5 frogs. They then write a sentence about adding 4 frogs and 5 frogs.

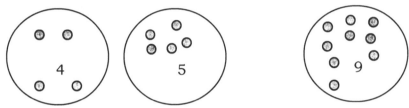

Reverse the above action for take-away. Place 9 frogs on a lily pad. Roll the die and take away that number. Write a sentence about the action.

STRATEGIES FOR MASTERY

Students need a variety of experiences in working with manipulatives and recording their own experiences prior to giving them strategies that will help them memorize their facts. Many times memorization is their first experience. When stress is placed on memorization, students might not develop an understanding of the concept. Without understanding, they will have a difficult time doing higher level thinking skills that are required in our technological age. The relationship between an addition and subtraction problem can be seen when we use the following terminology:

$$\begin{array}{r} \text{Addend} \\ +\ \underline{\text{Addend}} \\ \text{Sum} \end{array}$$

$$\begin{array}{r} \text{Sum} \\ -\ \underline{\text{Addend}} \\ \text{Addend} \end{array}$$

Students need practice using manipulatives prior to introducing them to the various strategies. Good teachers will build in the strategies as they are teaching the facts. They will also relate addition and subtraction. If students have internalized their basic addition facts, the subtraction facts are easy. They only have to reverse the operation, however. they must be helped to see the relationship.

Number families are again being used to help children see the relationship between addition and subtraction. A number family is all of the facts for a particular number. The number facts for the 7 family are:

7

7 + 0
6 + 1
5 + 2
3 + 4
7 - 0
7 - 6
7 - 5
7 - 4

and their commutative properties.

PROPERTIES OF ADDITION/ SUBTRACTION

Identity Element: Zero is the identity element for addition. By this we mean that anything added to zero remains that number. I + 0 = I

Commutative Property

Literature books will help children understand the commutative of numbers include Round Trip by Ann Jonas; Rosie's Walk; or Wheel Away by Ann Dodd. All three books begin at one point and move from place to place only to turn around and return to the original position.

This could be considered similar to reversibility. The commutative property means that 3 + 2 and 2 + 3 give us the same answer. This can be shown to children by writing the problem on a piece of oak tag, and using small clothespins to represent the problem. When the card is turned over, it will reveal 3 + 2 or the inverse of the number sentence.

We can look at the commutative property through the word commute - we leave home and go to school and we leave school, we go home. A good literature book to introduce this concept is "Round Trip:" by Ann Jonas. It shows how you can go one way and then turn around (and you actually turn the book upside down) and return to the original starting point. This reversibility eliminates almost 50% of the facts. If students do not understand the commutative property or the reversibility of a number, memorization or automaticity will be a burden, as they will have to memorize all 100 facts.

Some students will find their own strategies to learn the basic facts. Others will need to be helped to learn the strategies. It is suggested that the teacher involve the students in learning the strategies. The more they internalize the facts, the better they will be able to apply what they have learned.

An addition chart is in the back of the book (Appendix E). You might want to take it out and color code it to see how children learn their basic facts.

When Edward Thorndike gave us the right to use drill and practice as the only method of teaching mathematics, his work required the memorization of all 100 facts. He did not use the commutative property of the numbers. When we used the commutative, the number of facts to be memorized was cut in half. However, many teachers are still in the Thorndike era, and are asking their students to memorize all 100 facts.

ZERO FACTS

As mentioned before, zero is the identity element, and means that any number added to zero is that number. Time should be spent at this level, as students confuse the zero concepts when they begin multiplication. Look at the chart and color in all of the zero facts. If we use the commutative property, there are only 9 facts that need to be remembered. Without using the commutative property, there are 19 to remember.

+	*0*	1	2	3	4	5	6	7	8	9
0	0	1	2	3	4	5	6	7	8	9
1	1									
2	2									
3	3									
4	4									
5	5									
6	6									
7	7									
8	8									
9	9									

ONE FACTS

When one is added to the number, it is the next number. Activities that will help the students understand this is a number line. It is suggested that a number line be placed on the floor, and the students asked to move one step forward (addition) or one step backward (subtraction).

Other activities which will reinforce the one facts Is the use of the ladybugs or the frogs. Place "X" lady bugs on the flowers and one more comes - how many altogether? Place the frogs on the lily pads and one jumps off, how many are left? The students then write a story about their actions. When we use the commutative property, there are 7 facts that must be committed to memory. Without the commutative property there are 17 facts to be remembered.

+	*0*	1	2	3	4	5	6	7	8	9
0	0	1	2	3	4	5	6	7	8	9
1	1	2	3	4	5	6	7	8	9	10
2	2	3								
3	3	4								
4	4	5								
5	5	6								
6	6	7								
7	7	8								
8	8	9								
9	9	10								

TWO FACTS

The two facts can be reinforced in the same manner as the one facts. On the number line, they will jump 2 spaces instead of one. Be sure you start them at one and have them jump two spaces from one, landing on three.

+	0	1	2	3	4	5	6	7	8	9
0	0	1	2	3	4	5	6	7	8	9
1	1	2	3	4	5	6	7	8	9	10
2	2	3	4	5	8	7	8	9	10	11
3	3	4	5							
4	4	5	6							
5	5	6	7							
6	6	7	8							
7	7	8	9							
8	8	9	10							
9	9	10	11							

The counting on strategy can be used at this point. There are two developmental levels to use with the counting on strategy - the first is a lower developmental level where the student must count all of the numbers. E.g., if we are adding 3 + 2, the child would say "1, 2, 3", and then say "4, 5". If the child is at a higher developmental level, they will begin with the 3 and then count on to 4, 5. This is also an indication of the child's ability to gain speed with the facts. Those at the lower developmental level need more time and should not be forced to gain speed.

GAME: A game that can be played when you need to "eat up" minutes before recess or lunch or going home is "What is my numbers?". Give the student a number and they must give you 2 more than that number or 2 less than that number. This is also a good builder of mental mathematics skills.

By reinforcing the commutative property of numbers, we decrease the number of two facts that need to be remembered from 15 to 8. At this point by using the commutative property, we have reduced the facts to be remembered from 51 to 27.

DOUBLE NUMBERS

Double numbers are usually easy for children to learn. It could be the rhyme of the numbers, or the fascination of the use of doubles. There are 7 double numbers. Since double numbers to not have a commutative property, all seven must be remembered.

+	0	1	2	3	4	5	6	7	8	9
0	0	1	2	3	4	5	6	7	8	9
1	1	2	3	4	5	6	7	8	9	10
2	2	3	4	5	6	7	8	9	10	11
3	3	4	5	6	7	8	9	10	11	
4	4	5	6	7	8					
5	5	6	7	8		10				
6	6	7	8	9			12			
7	7	8	9	10				14		
8	8	9	10	11					16	
9	9	10	11	12						18

THREE FACTS

The Three facts can be discovered by the "counting on" method. The same developmental levels apply to the threes as to the twos. Using the commutative property, we reduce the number of facts from 12 to 6 that must be remembered.

NINE FACTS

The nine facts are interesting when trying to discover the pattern. **CAUTION**: children need to be at a higher developmental level in order to see and understand the patterns for the nines i.e., they result in a 2-digit number:
- the answer is one less than the number added to 9
- For example if we add 9 + 4 the answer begins with a 1, and the second number is one less than 4 or three. Therefore, the answer is 13.

+	0	1	2	3	4	5	6	7	8	9
0	0	1	2	3	4	5	6	7	8	9
1	1	2	3	4	5	6	7	8	9	10
2	2	3	4	5	6	7	8	9	10	11
3	3	4	5	6	7	8	9	10	11	12
4	4	5	6	7	8					13
5	5	6	7	8		10				14
6	6	7	8	9			12			15
7	7	8	9	10				14		16
8	8	9	10	11					16	17
9	9	10	11	12	13	14	15	16	17	18

Bridging to Ten: <u>An Alternative Method for the Nines</u>. Some teachers like to teach the children to bridge the 9's to 10 plus a number. We can discover our bridging to ten facts by allowing the students time to play with their manipulatives, and to practice making the changes. At first they will participate in a "hands-on" experience, and they will gradually begin to write their own sentences telling about their actions. Teddy Bears can be used for this activity; however, the ladybugs or frogs or any other manipulative can be used. The following is a story which was made up about a group of teddy bears.

> *One day some teddy bears were going on a train ride. The train master told them that the train would only move from the station if there were at least ten teddy bears in each car. The teddy bears got on the train and they waited and waited, but the train did not move. Pretty soon one of the teddy bears in the front car yelled *ten - then we could go on our ride.* Well all of the teddy bears were friends and wanted to ride together in the same car. So they sat and sat! Finally one teddy bear in the back car yelled to the front car "How many are there in the front car?' The front car answered "there are 9 of us.". The teddy bears in the back car talked about how many would have to go to the front car. They finally decided one would have to go before they could get their ride. One little teddy bear finally said he would move to the front car, As soon as he moved to the front car, the stationmaster said, " We have ten in the front car and 7 in the back car. Altogether we have 17, which is the same number as 9 in the front car and 8 in the back car. We now have the correct number in the front car, so we can go for our ride.' The teddy bears were very happy and they continued to ride on the train.*

In order for the train to move one teddy bear had to move to the front car in order to have 10 teddy bears in the car. Thus our algorithm now reads:

$$9 + 8 = 10 + 7 = 17$$

A train made out of popsicle or craft sticks, or boxes, could be made so the students could actually place the teddy bears in the car and move them around. This would be true for either the eight or the nine facts.

Once they have had a variety of experiences at this level, they can write about their actions. Their writing does not necessarily mean they have to tell the answer to the problem. It tells us they understand the methods to use in regrouping to 10's. Writing what they have done would help them understand their action i.e., there were 9 teddy bears in the back car and 8 teddy bears in the front car. I took "I" teddy bear from the back train and put it in the front train. Now there are 10 teddy bears in the front car and 7 teddy bears in the back car. Both had 17 in the cars.

Once they have had this experience they can look at the patterns for the 9's. The ones digit is one less than the number being added to the 9. Are there any other patterns that you can see on the chart for the nines?

There are 10 nine facts to be remembered, and five by using the commutative property.

HARD FACTS

There are some facts that are hard to find strategies for remembering

Naturally, if the students have had experience using manipulatives, these may not be hard. The hard facts are considered the fours, sixes, sevens and eights - 4 + 5, 4 + 6, 4 + 7, 4 + 8, 6 + 5, 6 + 7, 6 + 8, and 7 + 8. There are 8 of these facts. These facts can be worked on as a fact of the week. Each week give a fact that the students can take home. All week they are asked questions regarding the facts. Art projects, etc. could be integrated to help them feel they have internalized the facts.

+	0	1	2	3	4	5	6	7	8	9
0	0	1	2	3	4	5	6	7	8	9
1	1	2	3	4	5	6	7	8	9	10
2	2	3	4	5	6	7	8	9	10	11
3	3	4	5	6	7	8	9	10	11	12
4	4	5	6	7	8	9	10	11	12	13
5	5	6	7	8	9	10	11	12	13	14
6	6	7	8	9	10	11	12	13	14	15
7	7	8	9	10	11	12	13	14	15	16
8	8	9	10	11	12	13	14	15	16	17
9	9	10	11	12	13	14	15	16	17	18

Addition Charts: An addition chart can be made for each of the basic addition facts, e.g., the fives:

> 5
> 0 + 5
> 1 + 4
> 3 + 2

There are only 3 facts when we consider the commutative property of the five facts.

The addition chart can also be used for subtraction. If the want to find the answer to 5 – 4 = □ look on the five chart and find the 4. You discover the other fact is 1, therefore the answer was 1.

Remember, the use of manipulatives is a must for teaching children their basic facts. The basic addition and subtraction facts are to be mastered in the first grade. Some children will have a working knowledge of the facts, but may not be able to memorize them. Memorization comes with the developmental level of the child. If children in the grades after first do not know their basic facts, the teacher might want to revisit them and use manipulatives. After they have an understanding, then move to the various strategies. Different games will also help children learn their facts. They do not want to let their friends know they do not know their facts.

Don't hurry the learning of facts. Remember: rushing can cause "blurring" – and thus lead to learning problems.

Basic addition and subtraction facts are usually the focus of the first grade program. As the year progresses, double-digit addition and subtraction is usually added. However, many children are not developmentally ready for double-digit operations at first grade. The second-grade year usually focuses on double-digit operations in both addition and subtraction. Sometimes the basic multiplication facts are introduced toward the end of the second grade year. However, multiplication is the focus of the second and third grade classroom, and usually division is taught at the fourth grade level.

MULTIPLICATION AND DIVISION OF BASIC FACTS

What is multiplication?

"Multiplication is repeated addition". This is true, except it must be repeated addition of equal addends - or equal groupings. Multiplication is a shortcut to addition. If we have to add 9, nine times, we might have difficulty and even forget which numbers we had added. Multiplication allows us to compact the process.

Question: *What does a student need to know In order to learn multiplication and division?*

1. First, the student needs the same prerequisite skills as addition and subtraction.
2. Second, the student needs to know addition and subtraction (not memorization-a working knowledge is sufficient).
3. The student needs to know place value.

Question: *What stage of development are students in when they are learning multiplication and division?*

Most students are in a transition from preoperational to concrete operational. However, preoperational children can learn to multiply and divide.

Question: Where do we start in teaching the basic multiplication facts?

- Many times we can expand on the addition and subtraction skills to introduce multiplication and division.
- When we are teaching double numbers in addition, we could introduce 2 times a number. The learning situation at this time lends itself to multiplication.
- When we are talking about a single group, we could introduce the concept of I times a number.
- Division is a natural concept to introduce early since it is one of the first concepts children learn. They are always sharing with others. However, division is the last of the 4 basic operations to be taught.

Question: What skills can we use in the primary grades to introduce the concept of multiplication and division?

- Learning center activities can help reinforce the concept early.
- When we are dividing up materials, food, etc., reinforce the concept of division --- we had 48 pieces of candy --- how many did each student receive? 48 divided by "N" is?

Some activities that will help the students develop the concept of multiplication and division should be incorporated into the daily activities of the school day.

Multiplication - means "of"

If we look at 3 x 4, we can read it as what are 3 groups of 4? The question many times helps the student focus on the meaning of the algorithm.

Division - asks the questions "how many"

When we read the division sign. We ask, "how many 3's are in 24? In Multiplication we put things together. In division we take things apart.

The Geometry of multiplication; All multiplication facts must be either a square or a rectangle, Challenge the children to find one that isn't a square or a rectangle.

Manipulatives to teach multiplication

Problem 3 x 4 (means three groups of four)

Students place all of the unifix cubes together and count - or develop their own strategy.

Other possible uses of the problem -
Given 12 cubes, how many groups of 4 can you make?
Given 12 cubes, can you make three equal groups?

Cuisenaire Rods

The student builds a floor of "4 rods" until the "3 rods" will have an exact fit.

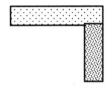

The students make a train out of the 4 rod and compare it to tens and ones. The student finds the answer is 12.

Division: Have the student make a train from a ten rod (orange) and a 2 rod. Find out how many 4 rods can ride on a 12 train?

Question: What 3 rods of the same color will be able to ride 12 train?

Both of these can be repeated for any multiplication or division basic fact.

Students are very excited to learn multiplication. If seems to be their ultimate goal - maybe they think they are 'grown up' once they can multiply. We need to do activities that will keep them excited about multiplication. Many times this is the point where students begin to be *turned off* on mathematics - the timed test tell them they are dumb.

Teachers who are willing to stray from the textbook and develop the concept before requiring timed tests will find their students will continue to be excited about multiplication because they understand what they are doing. If students understand. the concept the amount of memorization is lessened.

Again the California's Framework and the Principles and Standards for School Mathematics (NCTM) have stressed there is more to learning multiplication and division than memorization of basic facts.

Frog Multiplication

Literature Book: This section was first developed around a book called "Pond larker" by Fred Gwynne (Simon and Schuster Books for Young Readers, New York, New York). Other books about frogs can be used.

Activity 1: Pond larkers Children

Materials needed- *set of origami frogs or lima bean frogs, lily pads*

Directions: *Give the students a bag with origami frogs or lima bean frogs.*

STORY EXTENSION: Pond larker's *children went to the pond one day and saw some strange pink things on the pond. They ran home and asked Pond larker what these were and he told them he did not know. He advised the children to stay away from the pink things, but, as usual, one of his children could not stay away. The one child jumped into the lake and swam to a lily pad. When he reached the lily pad, he swam around the pink thing and decided to climb onto it. As he was resting on top of the pink thing the pink thing spoke to the one little froggy. It said, "If there were three on each lily pad they could float around the lake." The child called back and told those next to the pond that if three would climb on each lily pad they could sail around the pond. All of the other children immediately jumped in and swam toward a lily pad. The strangest thing happened. When more than three were on a lily pad, it would sink. They began counting to three. Once there were three on each pad, they would yell* **"STOP!"**... *Soon, all the pads were sailing around the lake. However, there were some sad little froggies standing next to the pond. Since there weren't three of them, they could not sail around the pond - the lily pads would only go when they had three frogs on them.*

NOTE: *The first time you use the story let them dramatize the actions, and then move to the manipulatives.*

After students have successfully completed this activity, give them some lily pads. and ask them to place 3 frogs on each lily pad. Ask them how many lily pads they had, and how many frogs did not have a lily pad. For example if they were given 13 frogs, they would have 4 lily pads with 3 frogs on each lily pad and one frog left over.

The first time the activity is completed, the teacher will record the number of lily pads, and the number of extras.

Number of Frogs	Number of Lily Pads	Number of Frogs on each Lily Pad	Extra Frogs
13	4	3	1
10	3	3	1
6	2	3	0

Extended Activities: The students use cards with one of the numbers from 1-6 written on each card. They will draw a card and that will tell them how many frogs will go on each lily pad. As they complete the activity, they will record the number of lily pads and the number of extras on a worksheet similar to the chart.

Activity 2: The Frog Express

Materials needed: *lima bean frogs, strips of 2-cm. paper*

Directions: *Problem Solving*:

The "Frog Express" is designing a frog car. The frogs would like your help in designing the types of cars they should build. For example, if they want to build a car to hold 12 frogs, what are the different combinations? (Note: the cars must be rectangular or square.) They could build 6 rows of 2 seats; 4 rows of 3 seats in each raw, 3 rows of 4 seats in each row, and 1 row of 12 seats.

After the students have designed the cars, they must use cm. Graph paper to build the train. They must also write an explanation of why they designed the car the way they did. If they are available, the lima bean frogs can be placed in the cars as a test.

Strategies for Learning the Basic Multiplication Facts

Again, as in addition, we may need to help some of the students learn the various strategies. Mastery is needed, but the students need to understand the facts first. Timed tests cannot be seen as the measurement of mastery.

Question: Can you do 100 multiplication facts in 2 minutes with 100% accuracy?

Activity 3: Equal Groupings

Vocabulary: *sets of. equal groupings, multiplication*

Materials needed: *manipulatives such as unifix cubes, colored blocks. centimeter cubes. breakfast food, buttons or any easy to handle object and colored circles made out of construction paper.*

Direct Instruction *(Remember to place the materials in a learning center for a period of free play) Have the students place some colored circles on their desk (e.g., 3). Direct them to place 2 of the manipulatives in each circle.*

Ask them how many they have altogether? The teacher can record the actions as 3 (draws a picture of 3 circles) and placing a 2 inside the circle. She then reads the action as "3 groups of 2 are 6" (gives the answer the students have given her.) This is repeated many times to give the students the idea of equal addends and the number of circles.

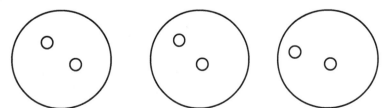

Relation to division: in reverse of this operation, give the students 3 circles and 6 faces. Ask them to divide the stars equally between the 3 circles? How many will each circle get?

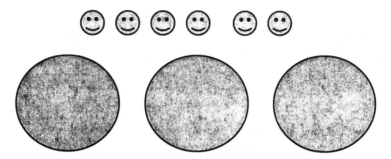

The teacher can also write 3 groups of 2. and then introduce the "x" sign as the words "groups of". When the concept is first introduced we want to use language the students can relate to. Some of our mathematical language has no meaning to young children. This process also helps our bilingual children relate to the concept being taught.

Learning Center Activities

Materials needed: Manipulatives in containers (several different containers can be used to give the students varied practice); colored circles; two dice (one with the number of *circles* and the other for the number or objects to be placed on each circle. You might want to use different colored die.

Directions: The student(s) choose a container and roll the two dice. If the circle die has a 3, they place 3 circles in front of them. If the number has a 2, they student places 2 object on each circle. The student records the following: 3 circles x 2 objects on each circle equal 6 altogether.

Arrays

Vocabulary: arrays, rows, columns

Materials: objects to make arrays, paper and pencil for recording arrays

Directed lessons: Ask the students to make an array of 4 rows and 3 columns. They will probably need help with this at first. Therefore, you might want to demonstrate on the overhead. Have the students place a circle on each intersection:

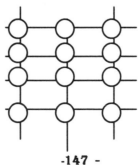

Question: How many rows do we have? (4)
How many columns? (3)
How many circles do we have altogether? (12)

The teacher records 4 x 3 = 12. The teacher then rotates the arrays one quarter of a turn and asks the students what this array shows (3 x 4).

Worksheets can be added to the above activities if you want to make a check on what the students are doing.

Strategies for multiplication and division

Multiplication like addition has many facts to learn if the *commutative property* of a number is not considered. When we use the commutative property, we reduce the facts by about 50%

ZERO

"Zero" is one of the hardest concepts that students learn. They confuse the zero in addition and the zero in multiplication. Students should have many opportunities to build zero facts, and then come to their own conclusion that zero times a number is zero. The book 2 x 2 = Boo by Lori Leady is a story that will help the students relate to the zero facts. If they are not learning the zero facts around Halloween, then make up stories to help them learn the facts.

X	0	1	2	3	4	5	6	7	8	9
0	0	0	0	0	0	0	0	0	0	0
1	0									
2	0									
3	0									
4	0									
5	0									
6	0									
7	0									
8	0									
9	0									

ONE FACTS

Again help the students discover that one times a number is that number

*One is the **identity** element of multiplication.*

TWO FACTS

The *two facts* are the double numbers in addition. We could teach multiplication by two at the same time we teach the double numbers.

X	0	1	2	3	4	5	6	7	8	9
0	0	0	0	0	0	0	0	0	0	0
1	0	1	2	3	4	5	6	7	8	9
2	0	2	4	6	8	10	12	14	16	18
3	0	3	6							
4	0	4	8							
5	0	5	10							
6	0	6	12							
7	0	7	14							
8	0	8	16							
9	0	9	18							

FIVE FACTS

The *five facts* relate to the clock time, and also to skip counting. Many times the five facts are easier to learn. To help introduce the 5's the teacher can use the clock. Beginning at the "1" and place a "5", next to the "2" place a 10, next to the "3" place a 15 ---on up to the 12 place the 60. This helps the students related to 5's in multiplication. This can be started as early as first grade.

Question: What is the pattern for the five facts

X	0	1	2	3	4	5	6	7	8	9
0	0	0	0	0	0	0	0	0	0	0
1	0	1	2	3	4	5	6	7	8	9
2	0	2	4	6	8	10	12	14	16	18
3	0	3	6	9		15				
4	0	4	8		16	20				
5	0	5	10	15	20	25	30	35	40	45
6	0	6	12			30	36			
7	0	7	14			35		49		
8	0	8	16			40			64	
9	0	9	18			45				81

DOUBLE NUMBERS

The *double numbers* in multiplication make a pattern - they make a square. To help students understand the double numbers, have them build the number with blocks. Help them discover that the double numbers make a perfect square. 4^2 means you must make the four into a square.

NINE FACTS

The nine facts are interesting. The answer to the nine facts must add up to 9. The tens digit is one less than the number being multiplied to the 9, and the ones digit is the number needed to add up to nine. There is also a finger multiplication that can be used with the nines. This is a fun game to teach to second graders. They are learning the nine facts without <u>realizing</u> they are learning their multiplication facts.

X	0	1	2	3	4	5	6	7	8	9
0	0	0	0	0	0	0	0	0	0	0
1	0	1	2	3	4	5	6	7	8	9
2	0	2	4	6	8	10	12	14	16	18
3	0	3	6	9	12	15	18	21	24	27
4	0	4	8	12	16	20	24	28	32	36
5	0	5	10	15	20	25	30	35	40	45
6	0	6	12	18	24	30	36	42	48	54
7	0	7	14	21	28	35	42	49	56	63
8	0	8	16	24	32	40	48	56	64	72
9	0	9	18	27	36	45	54	63	72	81

FINGER MULTIPLICATION

Although fingers can become a crutch in all basic operations, they can use used as a means of fun when learning the 9's facts. The nines are fun to teach. They are full of patterns, and we can also do some simple operations with fingers. Everyone has a different strategy for the nines.

Problem: 7 x 9

Some take 10 times a number and then subtract the number.

$$7 \times 10 = 70 - 7 = 63$$

Others take one less than the number multiplied by 9 and then figure out what they need to add to the number to get 9.

one less than 7 is 6   6+3=9

therefore the answer to the problem is 63.

Finger Multiplication 7 x 9

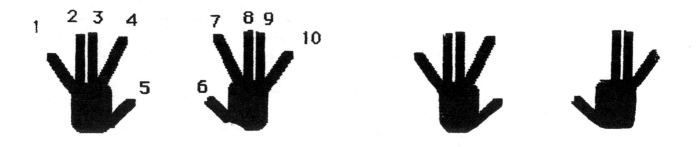

The fingers are numbered from 1 - 10. Since we are multiplying 7 x 9, we tuck the 7 finger under. The fingers on the left of the tucked down finger are tens, and those on the right are ones. Therefore our answer is 63. This will work only for the 9 facts. If the problem was 6 x 9, the sixth finger would be tucked down. There would be five fingers on the left of the six and four on the right, therefore, the answer would be 54.

OTHER FACTS

Once the students have had experiences with all of the multiplication facts, you might want to give them a multiplication chart. Assign them different multiples, such as 2, and have them color all the multiples of 2 red, 3 blue, etc. After they have completed this activity, see if they can recognize the attributes of the different multiples. This activity will help the students when they work with fractions. The term "multiples" will no longer be a strange term.

Hard Facts with Fingers. Finger multiplication for the hard facts can only be used for the numbers 6 x 6 through 10 x 10.

Problem: 7 x 8

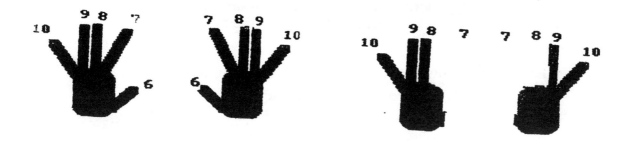

called "tens" and those extended are "ones". We know we have 5 tens. but we also know that 3 and 2 could be 5 if we added or 6 if we multiplied. Since this is a multiplication problem, we must multiply the 3 x 2. Therefore, we have 5 tens and 6 ones - Our answer is 56.

This method seems to cause confusion for many adults. Be sure you practice it and understand it before you attempt to teach it to your students.

Remember The fingers tucked down are tens, and we *multiply* the ones on the *left-hand* times the ones *on the right* hand.

LEARNING CENTER ACTIVITIES TO REINFORCE MULTIPLICATION AND DIVISION

To reinforce the concepts with Unifix cubes or beans or other manipulatives, the teacher can make index card sentences. The student constructs the sentence and records the answer on a worksheet similar to the following:

Multiplication: _____groups of _____

Division: How many groups of _____ are in _____?

Activity 1: Stars

Materials needed: 5 x 8 cards, colored gummed dots or stars (or star stamp), 2 die

Directions: *The students will roll the two different colored dice (or one color for rows and another column and then build an array with the model. If they roll 4 rows and 3 columns, they will use gummed objects to make the pattern. They will then write 2 multiplication and 2 division sentences to represent the pattern.*

Example: 4 x 3; 3 x 4; 12/3, 12/4

Activity 2: Unifix Cubes

Vocabulary: *groups of, altogether*

Materials needed: *unifix cubes*

Directed Instruction: *Students are directed to make four groups of three (the teacher is modeling this on the overhead projector). The students are then questioned about the number of groups and the members in each group (they have 4 groups of 3). The teacher records the sentence on the overhead or chalkboard. The students are then directed to tell how many they have altogether. This is accomplished either by counting, or by grouping into tens and ones. The teacher then records the total number. At this stage it is easier to continue with the terminology "groups of" instead of the multiplication sign.*

Activity 3: Cuisenaire Rod

> ***Vocabulary:*** *multiply, divide, equal*
>
> ***Materials needed:*** *Cuisenaire rods*
>
> ***Directed Practice:*** *Students are directed to take a 10 rod (orange) and see if any other colored rods of the same size could ride, on the 10 train.*
>
> ***Answer:*** *five red (2) rods, 10 white (1) rods, 3 Yellow (5) rods.*
>
> ***Reverse.*** *Take five red rods and make a train. What other rod is the same length of this rod? (orange)*
>
> ***Writing:*** *For both of these activities, the students are encouraged to write about their actions.*

Other Manipulatives

Egg Cartons.- The same procedures can be used with egg cartons. The egg carton is cut to have the correct number of cups. For example, If the student has 10 beans, or other manipulative, they are asked how many cups they will need to place two beans in each cup; 5 beans in each cup; and 1 bean in each cup.

Although these activities look like division, the writing of the actions reinforces the multiplication facts. The students can be directed to write 2 multiplication and 2 division sentences for each activity.

Two dice can be used - one to indicate the total number of beans, and the other the grouping needed. If a number comes up that does not give an equal division of beans, the students will have experience in division with remainders.

Basic facts are the foundation for all mathematics that children learn. Give them time, and make sure they have the self-confidence needed to learn all of the rest of their mathematics.

DON'T RUSH THE STUDENTS TO LEARN AND MEMORIZE THEIR BASIC FACTS. GIVE THEM TIME TO DEVELOP THEIR OWN SKILLS, THUS AVOIDING LEARNING DISABILITIES.

CHAPTER 11

1788 + 378 - 298 x 45
MULTIDIGIT OPERATIONS

Work with multidigit numbers is a marriage of basic facts and place value. If students are having difficulty with either concept, they may have difficulty with multidigit operations. Both the NCTM Standards and Principles and the California Framework list concepts with multidigit numbers as whole number operations - ranging from basic facts to double-digit facts. However, they do emphasize the use of problem solving as a method of reinforcing the multidigit operations.

Multidigit operations now begin as early as first grade, even though the students at this level may not be developmentally ready. Operations introduced at this level require a "hands-on" approach. Once they have had multiple experiences with manipulatives, they will begin to abstract the concept.

During the primary years, students need to have a variety of experiences with addition and subtraction. Subtraction is viewed as the "take away" model or the comparison. (Note: the comparison model can be very confusing to primary aged students. It is suggested the take away model be used for introduction, and until they understand the concept.) The missing addend is introduced (which can be the beginning of algebra), and multiplication and division begin to have meaning when they arise in real life situations.

The concepts of skills related to number and operations are a major emphasis of mathematics instruction to prekindergarten through grade2. When prekindergartners are asked how old they are, they hold up their fingers – something they have rotely been taught. As they grow, they develop meaning for the number of fingers they hold up. If asked to hold up three fingers, they can do it and relate it to other groups of three. It is the job of the teachers to help the children make sense out of number, and to introduce more sophisicated concepts as they move through the second grade. Second graders understand patterns, relationships, operations and place value.

As students move into grades 3-5, the concepts of multiplication and division of whole numbers are introduced. They still need manipulatives to introduce new concepts, but their knowledge base is such that they can now use diagrams, as well as the manipulatives. Students at this age are well aware of the relationship between the four operations. They can invent their own strategies and learn the properties of number. They also begin to add fractions and decimals to their knowledge base and are beginning to develop strategies for both whole and fractional numbers.

Understanding the operations with rational numbers is emphasized in grades 6-8. Students at this level continue to work with whole numbers in problem solving situations and continue to develop their sense of large numbers. Students at this level begin to revise their notations regarding the number system by moving into an expanded number system, i.e. positive and negative numbers, ratios, etc. They will also begin to develop proportional reasoning, which is needed for operations with fractions, decimals, rates, and ratios.

It is suggested that various computational tools be used in developing the operations – i.e. mental computations, estimation, paper and pencil, computers, and calculators.

ADDITION AND SUBTRACTION OF MULTIDIGIT NUMBERS

Question: What do the students need to know in order to add and subtract multidigit numbers?

1. working knowledge of basic facts (not memorization)
2. place value the position of the number determines its value
3. experience working with manipulatives to be used in the operations

Trends in Multidigit Operation

Recent research and trends give multidigit a new look. We used to structure the presentation of how we add and subtract numbers that require regrouping. We are now being told that we should give the children an algorithm, and let them solve it in any manner they wish. This is accompanied by their written explanation of how they reached their answer. Any solution is accepted that gives the correct answer. We assume they have the conceptual understanding of place value, and have had experience regrouping numbers.

As stated before once students have had guided instruction with the concept, they should spend at least 15 minutes a day in practice. A few problems correct are better than many problems wrong. Before moving to larger multidigit operations, be sure the students understand the 2 digits operations. If they are having difficulty here, stop and reteach.

Addition of Multidigit Numbers using Numeration Blocks

Multidigit operations begin in the primary grades with manipulatives. If numeration blocks are not available, the bean stick model can be used, or bean in a cup. All are acceptable models.

The first thing we want to do is to model the number of the problem. It is suggested that primary students make a tens and ones chart. An 8 ½ x 14 sheet of plain paper is all that is necessary.

We will give the students an addition problem.

Our problem is:

Tens	Ones

Model the first number 37

Model the second number

Tens	Ones
III	....
II	....

Collect all of the ones. (We have 11 ones)

We find that we have more than 10, so we will regroup and exchange 10 ones for one ten

Tens	Ones
III	
II	
	....

Once the exchange is made, physically carry the new ten to the top. (It is important to have the students do the physical action of carrying the new ten and placing it as you would in the written algorithm).

We then group all of the tens.
We find our answer is 6 tens and 1 one or 61.

Tens	Ones
—	
III	
II	.

If we let the students determine their own procedures, they may use something similar to this:

Tens	Ones
III II I	.

There is nothing wrong with this operation. It may take the child longer to complete his/her answer, however, as time goes on, s/he will find it easier to do the traditional methods.

```
              37
             +24
    30 + 20 = 50
     7 + 4 =  11

    50 + 11 = 61
```

ADDITION OF MULTIDIGIT NUMBERS USING A CHIP MODEL

When we built large numbers in our place value activities, we used a colored chip model. We are going to use this model for larger multidigit operations.

Materials needed: Colored Chips

When using colored chips for operations, it is suggested that the students have the algorithm next to them and record their actions. Again, this is to bridge the gap between the concrete and the abstract.

Step 1: Have students model the multidigit number as in the two-digit operations. For this model, the blue will represent the 100's, the red 10's, and the green 1's.

Problem: 378
+ 249

BLUE	RED	GREEN	
ooo (3)	ooooooo (7)	oooooooo	(8)
oo (2)	oooo (4)	ooooooooo	(9)

Step 2: Collect ones

As before, the students begin collecting the ones, then the tens, and then the hundreds. However, there would be nothing wrong with them collecting all of the ones, tens and hundreds, and then making an exchange.

BLUE	RED	GREEN
ooo	ooooooo	
oo	oooo	

ooooooooooooooooo (17 ones)

Step 3: Carry ten, record one

The student can now exchange the 10 yellow chips for a red chip, and records a "I" at the top of the tens column and a 7 in the ones column.

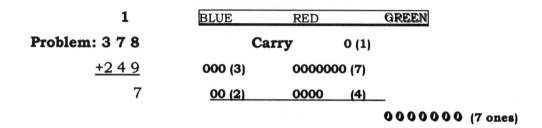

	BLUE	RED	GREEN
	1		
Problem: 3 7 8		Carry	0 (1)
+2 4 9	000 (3)	0000000 (7)	
7	00 (2)	0000 (4)	
			0 0 0 0 0 0 0 (7 ones)

Step 4: Collect the tens, and exchange for hundreds.

BLUE	RED	GREEN
000 (3)		
00 (2)		
(12 tens)	000000000000	0 0 0 0 0 0 0 (7 ones)

The student exchanges 10 red for I blue and records the one and the tens

		BLUE	RED	GREEN
1 1		(Carry) 0 (1)		
3 7 8		000 (3)		
2 4 9		00 (2)		
2 7				
		(2 tens) 00		0 0 0 0 0 0 0 (7 ones)

Step 5: Collect hundreds, record hundreds

	BLUE	RED	GREEN
1 1			
378			
+279	000000 (6)	00 (2)	0 0 0 0 0 0 (7)
627	(6 hundreds, 2 tens, 7 ones)		

The answer is 627. Students have used manipulatives and recorded their actions. Be sure they have an opportunity to practice these ideas. Let them use the manipulatives as long as they feel it is necessary. Students will decide when it is time to give up the manipulative, not the teacher, as they know when their visual images are replacing the actual objects.

NOTE: In the regular classroom, you would be introducing two-digit subtraction with manipulatives prior to the use of colored chips. However, in this section the addition is introduced followed by subtraction.

SUBTRACTION OF TWO-DIGIT NUMBERS

The steps modeled here are for the "take away" model of subtraction. Young children seem to grasp this method easier that a comparison model.

Step 1. The student models the two-digit number on the tens and ones chart (73), and add the digit cards to tell how many tens and ones are on the chart:

Problem: 73

    -36

Tens	Ones
IIIIIII	...

Step 2: The student reads the number to be taken away (36) and places the digit cards in the correct place. **(Do not model the 36)**

Since this is a take away model, the number being taken away is not modeled. It is written to help the student remember what is to be taken away.

Step 3: Question: "How many ones do your have?'
(Answer: 3)
Question: "How many ones do you have to take away?"
(Answer: 6)

When students are asked to take away 6 ones, they respond with "I can't". They are then asked how they can take away 6 ones. (Answer: regroup.) They are then asked to regroup.

Step 4: The student now regroups and has 6 tens and 13 ones, another name for 73.

The student is asked "Do you still have 7 tens and 3 ones?"

Since they do not, they must change the cards to reflect the regrouping and the action that has just taken place. A "six" card is placed on top of the "seven" card and a I is added to the 3 to make it read 13

Tens	Ones
IIIIII	...
	....
6 tens	13 ones

    6 13
    ~~7 3~~
   -3 6
   _____

Step 5: The student now takes away the 6 ones and places the remaining ones in the answer (7), and places that digit card in the column.

Tens	Ones
III	....
3 tens	7 ones

Step 6: The student now takes away the 3 tens and moves the remaining tens to the answer.

Step 7: The student reads the answer: 37.

Note: Subtraction is more difficult for children then addition, therefore, more time should be spent at this developmental stage. When students are doing worksheets, the addition and subtraction facts should be mixed, thereby forcing them to be aware of the attributes of the various signs. It is also suggested that two-digit addition and subtraction with regrouping be taught prior to non-regrouping. This forces the child to focus on the meaning of the numbers, rather than look at multidigit operations as an extension of the basic facts. In textbooks, you may have to skip pages that focus on non-regrouping and come back to them at a later time.

SUBTRACTION USING THE CHIP MODEL

Materials needed: chips, paper and pencil

PROBLEM: 5000
            -389 (Do not model this number)

Step 1: Model the 5000 using the chip model. The greens are thousands, blues are hundreds, reds are tens, and yellow are ones.

| BLACK(1000) | BLUE (100) | RED(10) | GREEN(1) |

ooooo

Next give them a number they are to take away (389). Again, we will be using a take away model. At this point, they can use the digit cards and put the correct number in each of the places, and also the number they will be taking away, or they can use paper and pencil to record their actions. We will assume they are using paper and pencil.

Since we have taught them to always take away the ones first, they will:

    go to the ones place (no chips)
    go to the tens place (no chips)
    go to the hundreds place (no chips)

Discussion: How can they get blue, red, and yellow chips?

The tendency is for the teacher to give them a short cut and just tell them to go to the thousands place or give them a pattern of 9, 9, 10. They don't let them discover their own pattern and method.

Step 2: Borrow and regroup from thousands to obtain hundreds and record action

Exchange 1 black for 10 blue.

```
  4 10
  5̶ 0 0
- 3 8 9
```

BLACK(1000)	BLUE (100)	RED(10)	GREEN(1)
0000 (4 thousands)	0000000000 (10 hundreds)		

Step 3: Borrow from hundreds to obtain tens, and record action

Exchange 1 blue for 10 red.

```
      9
  4 1̶0̶ 10
  5̶ 0  0
    3 8 9
```

BLACK(1000)	BLUE (100)	RED(10)	GREEN(1)
0000 4 thousands)	000000000 (9 hundreds)	0000000000 (10 tens)	

Step 4: Exchange 1 red for 10 green.

```
       9  9
  4 1̶0̶ 1̶0̶ 10
  5̶ 0  0
-    3  8  9
```

BLACK(1000)	BLUE (100)	RED(10)	GREEN(1)
0000 (4 thousands)	000000000 (9 hundreds	000000000 (9 tens)	0000000000 (10 ones)

Step 5: Take away from ones and record action.

BLACK(1000)	BLUE (100)	RED(10)	GREEN(1)
0000 (4 thousands)	000000000 (9 hundreds	000000000 (9 tens)	000000000 (10 ones)

Students now have 4 thousands (green), 9 blue (hundreds), 9 reds (tens), and 10 yellows (ones). They are now ready to begin their take away operation. Since all of the regrouping is completed, it does not matter if they take away from left to right or right to left. We are now ready to 'take away" the correct number in each case, and will use the right to left method.

```
       9  9
  4 1̶0̶ 1̶0̶ 10
  5̶ 0  0
-    3  8  9
  4  6  1  1 ones)
```

BLACK(1000)	BLUE (100)	RED(10)	GREEN(1)
0000 (4 thousands)	00000000 (6 hundreds)	0 (1 ten)	0 (1

Again, students need to have practice with using zeroes in different positions. Don't give them your shortcuts to make it easier. It may confuse them. Let them figure out their own shortcuts.

We need many exercises similar to this one in order for them to begin to pick up a pattern for exchanging.

Activities to reinforce addition and subtraction

Materials needed: Dot cards (see Place Value Numeration chapter), tens and ones chart, and manipulatives

Directions: The student turns over 2 digit cards and places them on the tens and ones chart. They model the number. The student then turns over 2 more digit cards and models the number on the cards. They proceed through the above mentioned steps. They can do this for several problems. After they seem to be able to do these, then they can go to the textbook and do problems given in the textbook. If the students don't need the manipulatives, let them do the problems without the manipulatives. However, if they make a mistake, ask them to use the manipulatives to show how they arrived at their answer.

Materials needed: Two dice with numbers 0-9 randomly placed on dice, paper and pencil

The new approach would take this problem:

$$\begin{array}{r} 33 \\ +28, \\ \hline \end{array}$$

The children would model the problem and then solve it without direct instruction. They would use a problem-solving approach: collect the tens collect the ones, and then exchange.

Their explanation might be: I collected all of the tens, and then I collected the ones. I had to exchange the ten ones for a ten, and then I put them altogether.

Another Controversy

There has been some controversy about direct instruction of multidigit operations. The old traditional method has focused on first teaching two-digit addition and subtraction without regrouping, followed by two-digit operations which require regrouping. Now some researchers feel multidigit numbers with regrouping should be introduced prior to multidigit operations without regrouping. The feeling is that operations with regrouping require the students to be aware of their actions, while non-regrouping is repeating basic facts with several numbers push?

Whichever method you choose, be sure it is developed with the use of manipulatives. Just because a student was successful with basic facts does not mean that student will understand place value. Students doing place value operations are probably at the concrete operation level.

Activities to Enhance Multidigit Operations

Palindromes: palindromes are numbers that can be read the same frontward and backward such as 434.

```
Directions:
Have the students write a 3-digit number          475
Reverse the number and add it to the first number +574
Add the numbers                                   1049
Reverse the answer                               + 9401
(Add the numbers)                                 10450
Reverse the number                                05401
(Add the numbers)                                 15851
```

The numbers read the same backwards and forwards therefore you have a palindrome. There are many words which are palindromes, e.g., dad, toot, etc.

MULTIDIGIT OPERATIONS FOR MULTIPLICATION AND DIVISION

Multidigit operations for multiplication and division follow much the same format and have the same prerequisite skills as addition and subtraction.

Multiplication

It is assumed that students have had experience with manipulatives when learning multiplication, and are able to see division as an inverse of multiplication. This understanding comes during the introduction of the basic multiplication and division facts.

When students are given a problem such as 4 x 17, they can use a proportional model to represent the numbers.

A discussion follows of the meaning of the "4". It means that the 17 is repeated 4 times, or 4 groups of 17.

After the students have modeled the 4 groups of 17 they are ready to collect the ones, regroup, and then collect the tens to find their answer. . They will need to regroup 2 groups of ten ones (turquoise and pink) to 2 groups of ten.

The students have discovered there are 6 groups of ten and 8 ones. 4 x 17 = 68.

The same procedure can be followed using a 3- or 4- digit number times a 1 digit number. However, at this time the <u>chips</u> should be used.

Problem: 5 x 376

The students will make five groups of 376. They will then combine the colors, exchange, and obtain a final answer.

As you can see, this operation requires several different steps of regrouping. However, it is a step that may be required for some students.

Expanded Notation

The grid method is another means of discovering the answer to the 2 digit number times a 2 digit number, i.e. 13 x 14. The students will need to be cautious when setting up the problem to make sure that only the corners are only touching.

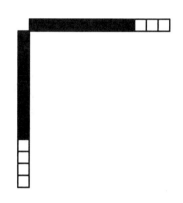

The student makes the grid and then fills it in as if it were a puzzle. They will need to match the width of the blocks in the original problem. They find a flat or 100 block is as wide as the orange ten rod. They will discover that a ten will be needed to match the size of the hundreds block. When all of these are filled in, they have some open spaces. Since multiplication is either a square or a rectangle, they will fill in with the 1's to complete the shape.

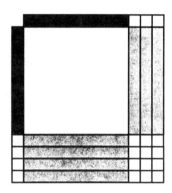

The next and very important step is to remove the original problem. This will eliminate the student' desire to include the original problem in with the answer. Once this is complete, they are ready to group the grid according to the values.

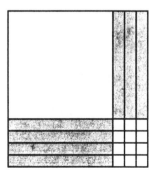

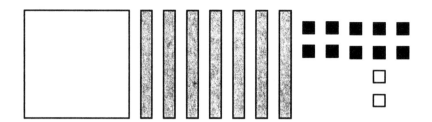

The students discover they have 1 hundred, 7 tens and 12 ones. This is regrouped to 1 hundred, 8 tens and 2 ones. Students have discovered that **13 x 14 is 182.**

There are other ways to do expanded notation. We will look at the problem:

```
    36
   x24
```

36	(30 + 6)
x 24	(20 + 4)
24	(4 x 6)
120	(4 x 30)
120	(20 x 6)
600	(30 x 20)
864	

This model uses expanded notation and solves it as four separate problems (Note: It can also be solved with either the chips or the numeration blocks.) Students must understand place value in order to complete a problem in this manner.

Distributive Method

Another method is to use the distributive property. Again, students need to have experience with this method before they can be expected to use it in practice. This time we have left the 36 as a whole number and distributed the 24 to 20 and 4.

```
36 x 20  +  36 x 4
 720    +   144  = 864
```

Variations of Distributive Method

In this case only the 24 is distributed, to 20 + 4

We take the 36 x 20 and then the 36 x 4. We then combine the two answers to give us a total answer.

Another method of doing the distributive method is to use expanded notation for both the 36 and 24. This is sometimes called the FOIL method which we use in Algebra.

```
(30 + 6)   x   (20 + 4)
(30x20) + (30 x 4) + (6 x 20) + (6 x 4)
  600   +   120    +   120    +   24   = 864
```

This requires higher level thinking on the part of the student.

A similar method to this is to have the students draw a box and fit the numbers into the box.

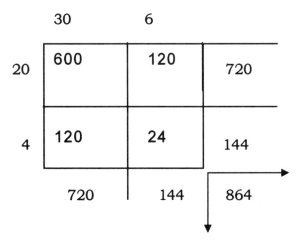

Lattice Multiplication (a method using Napier's Bones) is another method for helping students do multiplication with large numbers. This method only requires them to have knowledge of the basic multiplication facts.

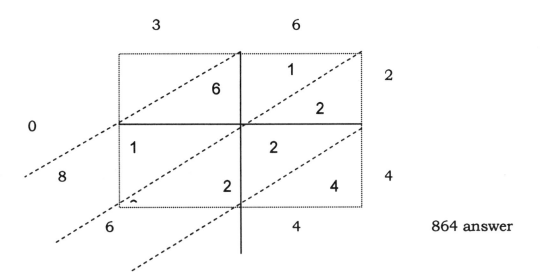

Note: to use the lattice method, you use simple multiplication. When your answer contains a ten and a one you place the tens above the line and the ones below the line. You then add on the diagonal. It is easier to place a card or piece of paper on the diagonal and then add. The answer is 0864 or 864. Notice the place value is in the correct position. If students have a tendency to add the 3, 6, 2, or 4, in with the answer, you might want to encourage them to cross out those numbers before they begin to add.

DIVISION OF MULTLIDIGIT NUMBERS

Division - division is the first concept where students start at the left and move to the right.

A good method to use to get students acquainted with the division algorithm, and reinforce that division is the inverse of multiplication, is the following method:

$$\begin{array}{r} 6 \\ 4 \overline{\smash{\big)}} \end{array}$$ (the student places the answer here)

Division is the first concept children experience but the last used in textbooks. Research says we introduce division by using the subtractive method (Greenwood) and they immediately change to the distributive.

Again when beginning to teach division, it is better to start with numbers the students can model and share. Give the students numeration blocks, or models of numeration blocks.

For example: if you have a group of 4 students, they will share 49 blocks. They can begin by sharing either the tens or the ones.

	Tens	Ones
Student 1:	1	2
Student 2:	1	2
Student 3:	1	2
Student 4:	1	2
Extras or left over		1

Each student will receive 1 ten and 2 ones, and there will be 1 left over.

By using this method they understand they are sharing 49 with four other students, and that each will get a given amount and they will have some left over.

After the students have had sufficient practice, and the teacher feels they have an understanding of the procedure, they then can eliminate three of the boxes, and the students should be able to visualize that all are sharing the same amount even though,. they are only seeing one answer.

Student 1,2,3,4

$$4 \overline{\smash{\big)}\, 49}^{\,12 \text{ with } 1 \text{ left over}}$$
$$5$$

The second problem should be a regrouping problem, e.g., 53. How can they divide 5 equally between 4 people? They find out each student will get I ten and they will have 1 ten left over. The next step is to exchange the 1 ten for 10 ones making a total of 12 ones. They then can divide the twelve ones equally among 4 people and move to the ones. They find they can give each child 3 ones, and have 1 left over. The answer they have is 13 r. I. This activity can be done using the period cards and digit numbers. The students will record as they manipulate their materials.

Once students understand the process of division, other algorithms can be used to help them move to our traditional algorithm.

Greenwood:

```
              31 r. 9
       24  | 753
             480      20(24)
             273
           - 240      10(24)
              33
              24       1(24)
               9    remainder
```

Distributive

```
        3                        31 r. 9
   24 | 753                  24 | 753
   -.  720                        720
        33                         33
                                   24
                                    9
```

Pyramid Method

Another method that has been used is called the pyramid method.

```
         1
        30
    24 | 753
        72
        33
        24
         9  Answer 31 R. 9
```

Once students understand the two-digit numbers, they are ready to move to multidigit numbers. Constantly assess the progress made by each student. If they seem to be having difficulty at any time, reteach immediately, before a major problem, occurs.

Note: See Appendix F for a historical review of checking strategies for multidigit numbers used throughout the ages.

Beware: don't force students to move faster than they are ready to move. If you rush them, they may develop learning problems.

CHAPTER 12

RATIONAL NUMBERS

Rational numbers are those numbers whose values could be less than one. Fractions, percentages, and decimals are the rational numbers that are part of the elementary curriculum.

What are rational numbers?

According to the new Webster's Dictionary, rational numbers are numbers that can be expressed exactly by the ratio of two integers. Rational numbers include fractions, decimals, ratios, proportions, and percentages.

The NAEP study of 1972-73 and 1977-78 found children were learning mathematical skills by rote memorization, and did not understand the underlying concepts. Both assessments found that 13 and 17-year old children successfully added fractions with like denominators, but only one-third of the 13-year- olds and two-thirds of the 17-year-olds could add I / 2 + I / 3. Only 24% of the nation's 13-year-old children were able to estimate the sum of 12 / 13 + 7 / 8 by selecting the correct answer among 1, 2, 19, 2 1. The fact that the two popular choices were 19 (28%) and 21 (27%) hints of the misconception.

Question: What do students need to know in order to learn rational numbers?

- Working knowledge of the basic operations
- Congruence
- Place Value (regrouping)

Vocabulary

Fraction: Fractions are a part of a set of rational numbers that can be expressed in the form A/ B when A is any whole number and B is any nonzero whole number such as 1 / 2. In this case A is a whole number one and B is a whole number, two, which is not a zero. Fractions have 3 parts – bar, number above bar, and number below bar. The word fraction means 'to break'.

Decimal: A decimal is a way of extending the Hindu-Arabic Numeration system to places less than 1. In decimals each place has a value of 10 times as great as the place to the right. Decimals have an implied numerator and denominator. The place of the last dot to the right in a decimal indicates the denominator.

Ratio: Ratios are a means of describing relationships between two or more groups.

Percentage: The term percentage means "out of 100" and the fraction is indicated.

Percentage and fraction skills need to be developed with the use of manipulatives. If students become involved with fractions at the concrete level, they will be able to develop an intuitive feeling for percentages.

Question: What stage of development are students in when learning rational numbers?

The stage of development will depend on when students were introduced to rational numbers. If students were introduced to rational numbers in Kindergarten, they were in the preoperational stage. They recognize the fractional parts, but may have difficulty with equivalent fractions if they are non-conservers. Equivalent fractions require students to regroup the fractional pieces just like they regrouped the digits in place value.

When students are beginning to do basic operations with fractions, finding the common denominator, etc., they are usually in the concrete operational stage and, in some cases, the formal operational stages. Decimals are usually introduced in the intermediate grades, where students should be at the concrete operational level.

FRACTIONS

Fractions are one of the most difficult concepts that students learn, yet one of the first concepts taught in the home. Fractions begin in the home as early as 2 years old. They are given a half of a piece of food, or are asked to share what they have with a friend or sibling. The question then remains: If it is taught in the home, why do we wait so long to introduce it in the schools? The second question is "Why is it so hard for students to understand when it has been a natural part of their lives?

Question: Why do adults have a difficult time dealing with fractions?

Maybe it is because they do not understand the true meaning of fractions, or maybe it is because they failed the sections on fractions in grade school. This lack of knowledge of fractions could again be another cause of students not going on to higher level mathematics, since most of the higher level mathematics is based on fractional functions.

Fractions need to be developed with the use of manipulatives. If students become involved with fractions at the concrete level, they will be able to develop an intuitive feeling for them. Division of fractions will be more than "invert and multiply".

What materials can be used to teach Fractions?

There are many materials that can be used to teach fractions - both commercial and teacher-made. When first introducing a fractional concept to students, one that usually leaves a lasting memory is food? When students eat their lesson, they can refer back to what they ate.

Once the students' interest in fractions is aroused, and they have had a chance to eat their fractions, other models can be introduced. Commercial models such as:

>design blocks
>fraction bars
>cuisenaire rods
>circle models, etc.

can be used.

The only precaution the teacher should take is to make sure s/he understands the process of using the manipulatives in teaching this concept. Teachers may have to go through the same process that the students go through during the lesson. Only when the teacher is confident with the manipulative, will the student be successful.

Teacher materials which can be used to teach fractions are:

>colored paper
>graph paper
>fraction strips
>circle pieces
>rectangular regions, etc.

Question: When should students begin the study of fractions?

The study of fractions should begin in Kindergarten and continue throughout school. All new fractional concepts should be introduced at the readiness level. A fifth grader needs readiness as well as a first grade student. If students have not had experiences with fractions, begin with developing an awareness of fractions - such as dividing the candy bars and tearing paper.

CONCEPT OF FRACTIONS

Part of a whole - first introduction at this level of order of fraction
Equivalent fractions
Part of a set (More difficult -use after region)
Basic operations with fraction models of fractions
Mixed number place value ratio

Readiness Activities for All Grade Levels

Part of the Whole

The first concept that students become aware of is the 'part of the whole'. It is easier to show equivalence with part of the whole than with part of a set. Candy bar fractions are an easy way to show the meaning of equivalent parts.

Candy Bar Fractions

Vocabulary: equal parts, unequal parts, divide

Materials needed: Two candy bars for each group (any candy bar that can be used). Students are allowed to ask questions, however, the only answer to a question will be "yes" or "no" – explanations are now allowed. When students decide which candy bar they would like a piece of, they will open the two candy bars. Usually there is some sort of rumbling regarding the ones in unequal pieces. The discussion follows that in order to be a fraction, it must be in equal pieces.

At the conclusion of the discussion, they get to divide the candy bars equally among their group and eat their lesson.

Once the concept of equal sized pieces has been established, then students are ready to further develop their understanding of fractions. An easy way to get them involved is to have them fold and tear fractions. All that is needed is colored sheets of paper (for young students use the whole piece of paper for tearing -for older students, strips of paper will work as well). This activity gives the students a manipulative that can be used at their desks.

PAPER TEARING FRACTIONS - students need to become involved in tearing fractions to gain an intuitive understanding that fractions are parts of the whole (colors can be changed according to what the teacher has on hand)

Activity 1: Paper Fractions

Vocabulary: *out of*

Directions: *Have the students take the white piece of paper and write "one out of one" on the sheet. Then move to the next color - have them fold the piece and tear (or cut) until all 6 pieces have been cut or torn.*

white - whole - write "1 out of 1"

blue 1/2 - write "1 out of 2" on each blue piece

pink 1/3 - write "1 out of 3" on each pink piece

yellow 1/4 -write "1 out of 4" on each yellow piece

green 1/6 - write "1 out of 6" on each green piece

orange 1/8 - write "1 out of 8" on each orange piece.

Note: *This procedure will probably take one class period.*

After students have written out the words 1 out of 2 on several pieces, the teacher may want to revert to the regular method of writing fractions, but keep reinforcing the terms "out of ".

As you are working with fractions, make the words 'out of' gradually to into a scribble and finally a straight line. This will help them read the straight line as the words "out of"

Step 1: On each fraction piece, write it as is shown.

```
1
out of
2
```

Step 2: On the next pieces write the word **Out Of** as shown

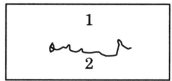

Step 3: Tell the students you are going to pull at each end of the word **Out Of and** make it a straight line

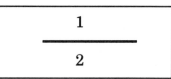

Step 4: The words **Out Of** are now straight lines. Reinforce the name of the line from time to time.

```
1
―――
2
```

After tearing the fraction pieces, students are ready to do operations with the pieces. As with whole numbers, one of the first and most important steps is to order and compare the fractions. Without this ability, they will not be able to recognize fractional parts, or use equivalent fractions, and will continue to think 1/ 12 Is larger than 1/ 2.

Ordering Fraction Ask the students to place the fractional pieces in order from largest to smallest or smallest to largest. As students order the fractions, they should be directed to recognize the difference in the size and the number written on each fraction - the larger the denominator the smaller the fraction. The teacher should record the sizes on the overhead or chalkboard.

Comparing Fractions: Ask the students to compare two fractions - 1 / 2 to 1 / 4. This is a good time to review equal, not equal, and greater than, less than which was introduced in the Basic Facts lesson. Students should have many hands on experiences when comparing the fractions. Again the teacher should record the actions in order to bridge the gap between the concrete and abstract. Once the teacher feels they understand comparing fractions, they can begin to do some paper and pencil activities with comparing fractions

1/3= 1/8   1/3 >1/8

This would be a good time to use fraction die or fraction digit cards. The fraction die would have simple fractions similar to those on the fractional sets. A set of digit cards could be more supplicated and use fractions such as 2/3 or 4/8. These cards can be made using half of an index card and writing the fractions on the cards.

Suggested activity 1: Cooperative Learning Activity - Place all of the fractional pieces in a pile, and student randomly picks a fraction. Student 2 randomly picks a fraction, and they write a sentence comparing the fractions.

Suggest activity 2: Give each student 2 dice (each labeled with the fractional parts that you have been working on in class). The student rolls the dice and writes a sentence comparing the two fractions. If they are unsure, they use their own fractional pieces to make the comparison.

Note: This activity can be used with students of all ages, especially those who do have a strong background in fractions.

Activity 2: Equivalent Fractions

Equivalent Fractions: One of the skills we are concerned about in fractional operations is an equivalent fraction. This generally is taught by telling students to multiply the top and bottom by the same number. However, if this skill is developed in the early grades though various activities, the student will gain an intuitive feeling for equivalent fractions. Work with equivalent fractions leads to converting unlike denominators to like denominators.

Vocabulary. Equivalent, another name

Directions: Ask the student to find a piece that is marked - one out of two. Ask them to find any other pieces that will fit exactly on the 1 out of 2 - they should find that 3 out of 6 fits and 4 out of 8 fits. This can be continued for each of the pieces. The teacher records "2 blue equal 1 white"; "2 yellow = 1 blue", etc. (This can begin in first grade.)

Bridging the-Gap

Worksheet suggestions

The first worksheets might want to look like this:

How many yellow = one blue?

    2 yellow =_____ blue

    3 pink = _____white

    4 orange = ----yellow

    or for older students: How many 1/4's in 1/ 2?   _____1/4 = 1/2

As they do more worksheets, eliminate the question:

    _____1/3= 1

Primary aged students can do these activities as long as they have the prerequisite skills, and have had lots of experiences with the manipulatives.

Students are actually beginning to divide fractions without knowing how to divide fractions or being aware of the rule "invert and multiply".

Other Readiness Activities

Activity 3: Paper Folding

Materials needed*: Whole sheets of paper, crayons*

Directions*: Take a whole piece of paper - fold it in half. Draw a line on the fold - Color 1/2 red. Fold the paper like it was, only fold it in half. Draw a line on the fold. Color ¼. What do you have? (1/2=2/4). Continue making other fractional parts. Color a part before drawing the next fraction*

Fold paper into 3 equal parts and draw a line on the fold

  - color 2 of the 3 blue

  - now refold the paper like you had it the first time.

  - fold it in half and draw a line on the new fold

What Fraction do you have now?

Activity 4: Partner Fractions

Materials needed: One set of fraction pieces

Directions: Give each pair of students a set of fractions. They are to record at least five ways of making equivalent fractions. (I / 6 + 1 / 6 = 1 / 3 or 2 greens = I pink).

Activity 5: Word Fractions

Materials needed: words

Directions: Write your name. Count the number of letters in your name. How many of the letters are vowels? What fraction of your name are vowels? The vowels are the numerator and the total number of letters is the denominator.

Extension: Can you find words that show the fraction 1/2 (half vowels and half consonants)? Try finding words that represent various fractions.

Have the students determine the fraction for their weekly spelling words.

Have the students determine the fraction for the names of each member of their family.

ADDITION AND SUBTRACTION OF FRACTIONS

Addition and subtraction of fractions can easily be taught with the use of manipulatives. The first operations will be with like denominators. The students will use their manipulatives to do the operations.

Activity 1: Basic operations with fractions

Vocabulary addition, subtraction

Directions: Directions similar to those listed below can be given:

Place 2 / 8's in your left hand

Place 3 / 8's in your right hand

How many eighths do you have altogether?

Record 2/8 + 3/8 = 5/8

You have 5 / 8's in your left hand.

Place 1/ 8 back on your desk.

What fraction is left?

Record: 5/8 - 1/8=4/8

Question: Can you find another name for 4 / 8 (a piece of paper where all 4 of the 8 pieces will exactly fit? (If needed, give them hints until they figure they can use a blue or 1/ 2)

Students should have many problems similar to this one when beginning to use fractions to do the basic operations. They can begin to make up their own addition and subtraction problems.

Suggested Activity: Students will use their fraction pieces and record addition and subtraction of fractions. The first activities should look at using one color only i.e. ¼ + ¼ = ½ .

As students understand the operations, they can use different colored pieces to cover a strip, i.e. **1/4 + 2/8 = 1/2**

Activity 3: Regional Fractional Models:

Materials: Fractional models

Directions: Give the students a fraction such as 3 / 4. Ask them to find ways they can express 3 / 4 as an addition problem.

Possible answers: 1 / 4 + 1 / 4 + 1 / 4

1/2+1/4

1/4 + 1/4 + 1/8 +1/8

At this point we are not concerned with like denominators. Our only interest is to have them find other ways of expressing the same fraction.

The same procedure can be used to have them write subtraction problems.

Ways to reinforce the concept

Cake: Give the students a cake (or bake one in class) and have them figure out how they can divide the cake equally among the members of the group. Next, give them two kinds of frosting and have them frost the cake according to the desires of the group. Write an addition or subtraction sentence regarding the way they frosted the cake. Eat the cake

Pizza: Give the students a round piece of paper. They can design a pizza according to their liking either by coloring the pizza or by finding pictures of ingredients in a magazine. After they have completed the pizza, have them decide how they will share their pizza. They then have to cut the pizza so that each person will get an equal sized piece (they must have more then 2 people – if they do not, have them invite a guest to share the pizza). This is also a good time to have them write a story about how they will share the pizza.

Games

Lotto

Materials needed: Approximately 14 cards with equivalent fractions

Directions: Shuffle the cards and place them face down between the two players. The players turn the cards over, one at a time and if they have two equivalent fractions, they keep the pair. The player with the most cards at the end of the game wins.

Bingo

Materials needed: plain piece of paper and markers

Direction: The students are directed to fold the paper into 16 squares (or however many the teacher wants). They are directed to trace the folded lines thereby making the squares for the bingo board. The teacher then writes 16 fractions on the board and the students decide which square they would like the fraction in. After they have all 16 fractions on their bingo card, the teacher calls out equivalent fractions and they match with their board. The winner is the one who gets 4 in a row.

Monster Fractions:

Materials needed, Set of equivalent fraction cards, one monster card

Directions: The game is played like Old Maid, only a Fraction Monster replaces the Old Maid.

MULTIPLICATION AND DIVISION OF FRACTIONS

Multiplication and division of fractions can begin in the primary grades without the use of the algorithm. If introduced at this stage, the algorithm will become another means of expressing an action. Again, we need to carefully develop the language with the operations. In this operation we will use the following:

Multiplication – "Of"

Question: What is 1/2 of 1/4

1/2 of 1/4
1/2 x 1/4

Division –"How many"

1/3 ÷1/6

Question: How many 1/6 are in 1/3 ?
or 1/3 contains how many 1/6 pieces?

The number sentence is an important part of the operation.

Activity 1: Multiplication of Fractions

<u>*Vocabulary*</u>: *of*

<u>*Materials*</u>: *Paper tearing fractions*

<u>*Directions*</u>: **Problem (I / 2 x 1/ 2)**

Take a 1 / 2 fractional piece (blue)
Fold the piece in half
What fractional piece will fit on half of the 1/ 2?
(A yellow piece or 1 / 4 should fit)

Problem: 1/4 x 1/2

Take a 1/ 2 fractional piece
Fold the piece in fourths
What fractional piece will fit on 1/4 of the 1/2?
(an orange piece or 1/ 8)

Problem: 1/2 x 2/3

Take two one-third pieces.
What is half of 2/3? (1/3)

Give the students problems similar to the above, but make sure they have the fractional pieces that will be the answers. (Problems can also be made that go with the fraction pieces in the Appendix G) Younger children can reinforce this concept by using number sentences, rather than the multiplication sign.

What is 1/2 of 1/4
What is 1/2 of 1/3?

Activity 2: Division of Fractions

Vocabulary: *How many*

Materials: *paper-tearing fractions*

Directions: *Problem: ½ / ¼*

Question: *How many ¼ pieces are in ½? Or*
½ contains how many ¼ pieces?
- *Take a 1/2 piece*
- *Take the 1/4 pieces*
- *How many 1/4 pieces will fit on 1/2? (2)*

Problem: 2/3 / 1/6
- *Take two 1/3 pieces (pink)*
- *How many 1/6 pieces will fit on the <u>two</u> 1/3 pieces? (4)*

Again, as in multiplication, only do problems which students can use manipulatives to get the answer.

Worksheets:

How many 1/4's are in 1/2? _____ 1/4 = 1/2

How many 1/3's are in 1? _____   1/3 = 1

Other materials that can be used in place of the rectangular region.

Fraction circles: Have sets of fraction circles available for the students to use in place of the squares. The same activities can be used as are used with the rectangular regions.

Activity 3: Fraction strips:

Vocabulary: *Fractional words*

Materials: *set of fraction strips*

Directions: *Students can have their own set of fraction strips (Appendix G)..*

They can cut them out and keep them in their desk for reference. Fraction strips can be made to include a wide variety of fractional pieces such as 1/20, 1/24, etc. Students do comparing, adding, subtracting, multiplying and dividing of the fractions.

Students should have had many experiences with equivalent fractions by this time. They should have not only manipulated like fractions, but recorded the manipulation, so they have some idea that there are many different names for the same fraction.

Activity 4: Unlike Denominators

Vocabulary: *Denominator, like, unlike*

Materials: *Fraction pieces*

Directions: Direct the student to take out a 1/2 fraction piece. See how many other names they can find for the 1/2 piece. (3/6. 2/4. 4/8). The teacher records all the names for 1/2 on the board. After students have reviewed this skill, then they are ready to do some simple adding with unlike denominators.

Problem 1/2 + 1/4

Method 1: Measurement

Place the 1/2 and 1/4 together

Find out which equal sized pieces will fit on the region.

They may find out that three 1/4 pieces fit. Or even six 1/8 pieces. If they say six 1/8's, have them find a larger group of pieces that will fit on the region.

They should find out that three 1/4 pieces fit.

Method 2: Exchange

Place the 1/2 and 1/4 together

Exchange the 1/2 for two 1/4 pieces

(Since all pieces should be the same size, they will have to **exchange** the 1/2 for two 1/4 pieces)

Add the three 1/4 pieces and get 3/4.

Some students will automatically say the answer is 3/4 – they have already visually exchanged the 1/2 for two 1/4 pieces and added.)

Subtraction problem: 1/4 – 1/8.

Method 1: Measurement

Take a 1/4 piece and place a 1/8 piece on top. What piece is needed to complete the puzzle? – 1/8

Method 2: Take away

How can they take away one eighth from one fourth?

Since the denominators are not alike (and the colors are different), something has to be done to make all colors alike – since 1/8 is to be taken away, the student should regroup to get eighths (all colors are then the same), if possible. In this case the student can regroup a 1/4 to 2/8

Take away 1/8 and there is 1/8 left.

The above activities are suitable for any grade level, but the emphasis is placed on the primary grades. Once students are older and are beginning to develop some logical thinking skills, they will be able to do and understand advanced operations.

FRACTION STRIPS

All of the above activities can be used with fraction strips. Older students seem to prefer these to paper tearing and coloring. Each student can make their own set of fraction strips, or they can be available for students to use.

It is suggested that the strips be glued to poster board or art board to make them easier to handle. The following activities are suggested for students in the intermediate grades.

Activity 5: Fraction Strips

Materials needed: *fraction strips*

Directions: *As suggested before, one of the first activities that students need to do is to compare and order the fractions.*

Worksheet 1: Order all of the fraction strips, and listing in order from smallest to largest

Worksheet 2: Give each student 2 dice with fractions written on them.

Students roll the dice and write a sentence comparing the two fractions.

Worksheet 3: Give the students 5 fractions: Have them write at least three equivalent fractions for each given fraction.

Worksheet 4: Start with simple addition and subtraction problems with like denominators. Have the students complete the problems, and then give their paper to a partner. The partner must model the problem to check the answer. Any wrong answers must be corrected.

Worksheet 5: Addition and Subtraction problems with unlike denominators. This time the emphasis of the worksheet will be on unlike denominators. The students will only find the like denominator – not complete the problem. The next day they will complete the problems, once they are sure they have the correct denominator.

Worksheet 6: Multiplication of fractions:

Again, at this time the word **"of"** will be used. The reason for this is to allow them to manipulate the material rather than concentrate on the sign. This procedure will require them to do some thinking. For example, in looking for the answer to "What are 1 / 2 of 1/4?" – they first must find what two fractional pieces equal 1 / 4 (some may have to try several pieces before coming up with two 1 / 8 pieces), and then find the answer of 1/8.

Worksheet 7: Division of fractions:

As with the multiplication, this exercise should focus on the material, not the written problem. Question: How many 1 / 4's are in 1 / 2?

This seems easy, but they need practice at this level prior to working with larger fractions or with the rules we impose on them.

Worksheet 8: Mixed numbers.

A common misconception is that fractions must be less than a whole. Have the students do some problems with the fraction strips. Be sure you have several whole strips for them to use.

>Addition: 3 1/4+1/2=

This requires regrouping plus finding the common denominator.

>Subtraction: 2 5 / 8 – 7 / 8

This problem requires the student to regroup or exchange 1 whole for 8 / 8 before doing the subtraction problem.

>Multiplication: 1/2 X 2 1/4

>The question is: What is I / 2 of 2 1 / 4

The student will have to model the 2 1 / 4 and then find out what I / 2's of it is

>Division: 4 1 / 4 ÷ I / 2

>The question is:" How many I / 2's are in 4 1 / 4?"

The procedure here is easier, as the student can measure the number of I / 2's that are as long as 4 1/4.

As students progress in their work and seem to understand multiplication and division, then the sign for each can be used. Once the sign is used, they should continue to use the manipulatives, and the teacher should help them discover the pattern for multiplication and division without telling them.

PATTERN OF MULTIPLICATION

Question: What is the pattern for multiplication?

>1/2 x 2/4 = 2/8
>1/3 x 4/8 = 4/24

Students should be able to tell the pattern:

>the numerator times the numerator;
>the denominator times the denominator.

When students first begin this operation, they will probably forget to reduce or find another name for the fraction. Don't count their answers wrong – question their answer and see if they can come back with a lower fractional name. Again they must be able to find the pattern in order to reduce fractions to their lowest form.

Problem: ½ x ¼

Draw a rectangle and shade in 1 / 2 of it with yellow. Divide it into 1 / 4 pieces and place x's on 1/4 of it.

Questions: How many squares are both yellow and have an "x" on them?

The answer is one out of eight or 1/8.

Division of Fractions

Introduce the division sign in place of the questions "how many" There are several ways to use the division sign. It can be introduced in a way similar to the division of whole numbers.

1.   1/4 ⌐1/2

2.   1/2 / 1/4

   4 1/2
      ―――
      1/4

Whichever method you choose, be consistent and introduce it carefully. No matter which algorithm you choose, the question remains "How many 1 / 4's are in I / 2?". At this point, we have not yet talked about inverting and multiplying.

PATTERN OF DIVISION:

First introduce the concept of the reciprocal - the opposite of the denominator. Forms 2 and 3 are probably easier to use to show this operation.

   4   We want to get rid of the denominator

Question: *What can we do?*

Introduce the reciprocal of 1/4. If we take 4 / I times the denominator, we must also take it times the numerator.

$$\frac{1 \times 4 \; 2}{2 \times 1}$$
$$\frac{1 \times 4}{4 \times 1}$$

In this case we can cancel out the 4's and the 1's in the denominator and we still have a 1 left. The numerator cancels out the 1's and you have 2/ 1.

The problem now reads

$$\frac{\frac{2}{1}}{1}$$

Note we still have a denominator of 1 – we did not get rid of it completely.

To complete the problem we now have 2/1 or an answer of 2

This tells us there are two 1/4's in 1/2

Games and activities using the fraction strips can help reinforce the concepts and help students develop a visual representation for fractions.

Working with Large Fractional Numbers

We need some help in working with fractions with large uncommon denominators - in fact some of the numbers seem impossible. Prime and Composite numbers are useful at this stage.

Vocabulary

Prime Numbers: numbers whose only factors are themselves and the only way we can multiply to get 13 is 1 x 13 or 13 x 1. There are many prime numbers.

Composite Numbers: -all other numbers that are not prime. They are numbers that can be broken down into equal groupings. For instance 15 can be:

5 x 3
3 x 5
1 x 15
15 x 1

Factor: numbers that we multiply times each other to get a larger number. In 15, the factors are 1, 3, 5, and 15. We multiply sets of these factors together to get 15.

Multiples: multiples are like skip counting - the multiples of 15 are 15, 30, 45, 60, etc.

(Maybe we should concentrate on just multiples and factors for a period of time before we begin finding the common denominator. The language we use in connection with multiples and factors are confusing. We want the least common multiple: least means smallest, while multiple indicates biggest. The same goes for the Greatest Common Factor - greatest means largest while factor means smallest. No wonder everyone gets confused.)

Question: What good are prime and composite numbers?

Prime and composite numbers allow us to break large numbers down into smaller workable numbers. We find the common factors to accomplish this. One method is to use "factor trees".

Problem: 6/15 + 7/12

Since 15 and 12 have unlike denominators, we must find a number they both have in common. or the Least Common multiple (i.e., the smallest number they both have in common).

FACTOR TREES

Factor trees are one method of finding the LCM. We always start with the smallest prime number, 2. Since 15 is not a multiple of 2 (no matter how long we skip counted by 2's, we would never reach 15), then we must try 3.

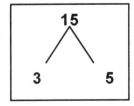

We find that three is a factor of 15 and we know (from our experience with multiplication) 3 x 5 = 15.

12 is a multiple of 2 therefore we can use the 2.
We know 2 x 6 = 12 so we include that in the tree.
Since 6 Is not a prime number, we must break 6 into
2 x 2 x 3, which are prime numbers?

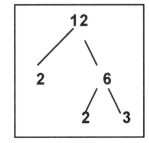

The prime factors of 15 are: 3 and 5.

The prime factors of 12 are: 2, 2, and 3

If we multiply all of these prime factors together we will get 180. That is what we would get if we multiplied 15 x 12. If that is the case, then why go to all of the trouble with factoring? We do not need to use all of these factors since some are common to both 15 and 12. Therefore a chart will help us find out which ones we need

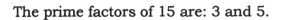

We notice they both have a 3 in common; therefore, we will only use one of the 3's. We now can multiply 2 x 2 x 3 x 5 and get 60 as the least common multiple. We are now ready to complete the problem.

Other methods for finding the common denominator or Least <u>Common</u> Multiple.

Remember: Don't choose a method that makes sense to you; it must make sense to the student.

Skip Counting

Another method that can be used in doing this is to skip count.

Use the calculator and record the multiples of each number.

USING CALCULATOR TO FIND COMMON DENOMINATOR

12: press 12 + = and record the number each time you push the equal sign

Do the same with 15

        12 24 36 48 **60** 72

        15   30 45 **60** 75.

We check and find that both have the number 60 in common, therefore, 60 is the common denominator.

VENN <u>DIAGRAM</u>

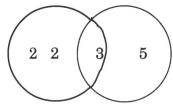

Again we see that the factors are 2, 2, 3, 5.

RENAME FRACTIONS - USING A MULTIPLICATION CHART

The multiplication chart is an easy reference to equivalent fractions.

X	1	2	3	4	5	6	7	8	9
1	1	2	3	4	5	6	7	8	9
2	2	4	6	8	10	12	14	16	18
3	3	6	9	12	15	18	21	24	27
4	4	8	12	16	20	24	28	32	36
5	5	10	15	20	25	30	35	40	45
6	6	12	18	24	30	36	42	48	54
7	7	14	21	28	35	42	49	56	63
8	8	16	24	32	40	48	56	64	72
9	9	18	27	36	45	54	63	72	81

If we want to find equivalent fractions for 2/7, we cover all of the numbers except 2 and 7. We see that 4/14, 6/21, 8/28, 10/35, etc. are all names for 2/7.

2	2	4	6	8	10	12	14	16	18
7	7	14	21	28	35	42	49	56	63

We can do the same for 8/9.

8	8	16	24	32	40	48	56	64	72
9	9	18	27	36	45	54	63	72	81

Students can be given a multiplication chart to fill out. Once they fill it our, they can cut apart each number. This will enable them to put two equivalent fractions together to discover the various names.

If we are looking for other names for 2 / 3; place the two multiplication factors on top of the three multiplication facts and read across the line. If we are looking for equivalent fractions for 1/ 9, place the one on top of the nine and find new names for 1/9.

Worksheet: Give the students two numbers and have them list the equivalent fractions for those numbers.

Question: *What is the pattern?*

Change 2/3 to 8/12 -instead of telling the student to multiply the same number times the top and the bottom, let them discover how they got from 2 to 8 and from 3 to 12.

ACTIVITIES FOR FRACTIONS

What's the Whole?

Give students a 30-inch strip of paper. Tell them they can divide the strip into equal parts (no more than 10 equal parts. After they have made their fraction, have them tear or cut off part of the fraction (be sure to tear on the fold line) For example a child divides his / her strip into 5ths and tears off 2 / 5's. The child writes "two out of five" on the strip that is given to the partner. The partner must then make a piece that represents 3 out of 5 and return it to the owner. The owner then checks to see if the partner is correct.

This could also be used in random fashion to make a whole. The children then write a story about their fractions. "Ben gave Jair 2 / 5 of his piece and Jair added 3 more fifths to make five fifths which was the original fraction.

GAMES FOR FRACTIONS

Materials needed: Worksheet two 0-5 number cubes two different colored pencils or crayons

Directions: The person rolling the highest number starts the game.

1. Lead player rolls both cubes. (If a zero is rolled the player loses that turn.)

2. The player thinks of the larger number as parts of a circle on the worksheet.

    The smaller number is the parts of the circle that they are to color. (I.e. if a player rolls a 4 and a 5, the 5 becomes the denominator and the 4 the numerator. The player will then color the circle that shows 5ths. They may choose to color one fifth of one circle and three fifths of another circle, etc.

3. The player initializes the part of the circle they have colored. After all circles are colored, the person with the most parts colored wins that circle. If each one colored half of the circle, no one wins that circle.

4. Play continues until all circles are colored. The one with the most circles colored wins the game.

Make One

Materials: Deck of cards consisting of:

    2 cards for each of the following fractions (1/2, 2/4, 2/8, 4/8, 6/8)

    4 cards for each of the following fractions (I / 4, 1/ 3, 1 / 4, 1 / 8, 3/8, 5 / 8)

Directions: Each player is dealt 5 cards and one card is turned face up on the table. The object of the game is to make as many suits of cards that "Make one". The first player sees if he has a card in his/her hand that, when added to the turned up card, makes one. If s/he does not, s/he "adds on" to the card showing, by putting out one of his cards, so the sum is less than one. S/he then picks a card from the stack. Each player proceeds in like manner. The player who successfully "Makes One" takes in the suit of cards. If a player has no card to play he picks a card and then passes.

The player with the most units at the end of the game is the winner.

Fraction

Materials needed: Fractions cards for 2 players (Any fractions can be used)

Directions: Deal out the entire deck, face down. Each child turns up one card. The player with the higher cards takes and places them both at the bottom of his deck. If both cards are the same (or equivalent) a second card is played, face down, on top of the first. A third card is then played; face up, on top of the second. The highest card wins. Play continues until one player has won all the cards or a specified time limit has elapsed.

Old Maid

Materials needed: Cards with fractions, decimals, percents or mixed operations. (Be sure you have matching sets (i.e., 3 / 5 and .60) if using mixed operations) Create your own type of old maid or monster Procedure: Play the same as you do the regular OLD MAID card games.

Bingo

Materials needed: Piece of plain paper for each student, fractional numbers (decimals, percents) for the board

Directions: Have students fold the paper into a given number of squares. Have a free square in the middle. Write specified numbers on the board and tell the students that they can be placed anywhere on their paper. After everyone has their own board made, play proceeds as with regular bingo.

ONCE STUDENTS HAVE A FAIRLY GOOD GRASP OF THE REGIONAL USE OF FRACTIONS, THEN THE USE OF SETS SHOULD BE INTRODUCED.

Parts of a Set

Materials: Small package of M & M's or other candy that can be divided into equal sized groups

Directions: Have the students count the number of M & M's in the group. (At this point you might want to make sure that all of the students have the same number of M & M's. If not, have them share, so each group will have the same number.)

Step 1: Have them divide the candy equally with the members of the group. This will give the teacher the opportunity to write a problem on the board - for example:

Each student is given 1 / 4 of a bag of M & M's

$$1/4 \text{ of } 40$$
$$1/4 \times 40 = 10$$

Each student received 10 M & M's: 1/4 of 40 is 10.

Step 2: Give them the following problems

$$2/5 \times 40$$
$$1/5 \times 40$$
$$1/2 \times 40$$

Have them use the M & M's to determine the answers.

Activities that have been used for parts of a whole can be adapted to use for parts of the set.

AT THIS POINT STUDENTS NEED TO BE AWARE OF THE RATIO OF THE NUMBERS. Ratios are another way of expressing fractions. The use of the M & M's is another good way to express ratio.

Ratios

Ratios are a means of describing relationships between two or more groups. For example, what is the relationship between red M & M's and the whole set of M & M's – (brown, green, yellow, etc.)

If there are 7 red M & M's in the whole bag, then the ratio would be 7: 40.

Worksheet: Express the ratio for each color of M & M's.

Question: Why do we have so many different ratios for the same color?

The ratio for two numbers can be written as a fraction, i.e., 5:1 for five toes to one foot.

Classroom Activities for ratio

Compare boys to girls,

blue eyes to green eyes,

centimeters to meters,

Saturdays to number of days in a week,

pennies to nickels.

RELATING FRACTIONS TO DECIMALS

Although decimals were introduced with numeration, we have not related them to our numeration system. Decimals are probably one of the most important concepts that we can teach since our momentary system is built on the use of decimals. It has been stated that many times students who are new to this country relate to money as the use of decimals. It is something concrete that they use each and every day.

A decimal is a way of extending the Hindu-Arabic numeration system to places less than 1. In decimals, each place has a value of 10 times as great as the place to the right. Decimals have implied numerator and denominator. The place of the last digit to the right in a decimal indicates the denominator.

Activity 1: Fractions are decimals and decimals are percents

Vocabulary: *fraction, ratio, percent*

Materials: *Several pieces of 10 square x 10 square paper, colored pencils, crayons, etc.*

Directions: *Have the student fold their square paper in fourths and draw a line on the folds. Direct them to color one fourth of the square. Write 1/4 on the board or overhead*

Questions: *How many squares are on the paper? (100)*

    How many squares are colored? (25)

    What is the ratio? (25: 100)

    How can this be expressed as a fraction? (25 / 100)

Introduce the concept of decimal - less than a whole

    Is this a whole or part of a whole?

The decimal means part of a whole

Question: *What is another name for 25/ 100? (1 /4)*

    Introduce the concept of percentage – "out of 100"

    What percent is colored? 25%

At this time a chart could be established:

Fraction	Ratio	Decimal	Percent
1/2	50:100	.50	50%

Other problems should be completed similar to the above. The more time spent at this level, the more students will develop an intuitive feeling for the relationship between fractions, ratios, decimals and percents. The decimal /fraction/ percentage charts will help students see the relationship. You might want to give the students problems where the fraction is given and have them find the decimal or percentage.

Again as in the other games and activities, various games can be made to help the students see the relationship. Bingo or lotto games are simple and easy.

Other activities that can be used for showing the relationship:

Give students a 10 x 10 array without lines. Give them a decimal fraction and have them fill it in with numeration blocks. For example: .2   .35

Have them express the decimals as fractions, ratios, decimals, and percents.

MONEY

Students can be given play money to use in different decimal operations. Rectangular pieces can be given to students and they can design their own dollar bills. Circles can be colored to represent pennies and dimes at this time we do not want to use nickels and quarters as we are talking about the base ten system.

It is said that the operations with decimals are identical to those with whole numbers, except that a decimal point is added. Some have actually suggested that students be taught to ignore the decimal sign when doing the problem, and then going back and placing the decimal.

Carpenter et al in 1981 reported a lack of conceptual understanding about decimals in 9 and 13 year olds. He advocated building a strong understanding of the decimal concept before proceeding to computation and application. He suggests doing this by either building on whole numbers, or fractions.

When students were asked which was larger .37 or .73, the students stated that .37 was larger because it had 3 tens and 7 ones. What is the misconception?

Problem: .26 +. 49

Students can ignore the decimal points and add 26 and 49. The answer would be 75, and since we have two decimal places to consider, we would start at the right side of the number and move two places to the left, thus the answer would be .75 or seventh five hundredths.

Concrete Level:

"Model 2 tenths and 6 hundredths. Add 4 tenths and 9 hundredths." Regroup what is your answer?

Pictorial Level:

Graph paper.

Color 2 tenths and 6 hundredths (I color)

Color 4 tenths and 9 hundredths

How many are colored all together.

(a different color)

    Some students have difficulty keeping their decimals in a row. A double bar can be drawn between the whole number and the decimal and the decimal must always go in the double line.

Multiplication of decimal.

Cross multiplication – "What do they both share?" .2 x .3

.2 x .3

How many decimal places are to the right of the ones place in the first number? (1)

How many decimal places are to the right of the ones place in the second number? (1)

How many decimal places are to the right of the ones place in both numbers? Since there are 2 places we must start at the right and count 2 places to the left and place the decimal point. In this instance, there is only one number; therefore, a zero must be used as a placeholder.

$$\begin{array}{r} .2 \\ \times .3 \\ \hline \mathbf{.06} \end{array}$$

Division of Decimals

When dividing decimals we have learned that what we do to one side, we must do t the other side.

$$.54 \overline{\smash{)}7.89}$$

In following the rule we would move the decimal point two places to the right, and then proceed with the problem. Why do we move it two places to the right?

This should be taught as: $.54 \times 100 \overline{\smash{)}7.89 \times 100}$

We multiply each number x 100 in order to remove the decimal point. Our new problem will not read:

$$54 \overline{\smash{)}789}$$

We then proceed as we would for any division problem.

Hint: When writing decimals, in order to keep them in the correct order, have children draw a double line for the decimal. The decimal point must always go in this line. The ones place is the focal point for decimals - going to left or right the decimals have the same place value name - the only difference is when the number is less than one It is expressed with a "th'

thousands hundreds tens ONES tenths hundredths thousandth

Card games such as War, Go Fish, Old Maid, etc. can be used to reinforce the relationship between fractions and decimals.

War: Cards could read 1 / 2 and .5

Go Fish: I / 2, .5, 4 / 8,

Old Maid: Use the same deck as War only add some type of Old Maid card.

The card games could be adapted to include all three operations - fractions, decimals and percentages.

Probability

Probability can be described as a division problem "f/ n?" where f is the number of favorable outcomes and n is the total number of possible outcomes

What is the probability of drawing a red card from a deck of cards?

First we know that there are 52 cards in a deck, and half of those cards are red. Therefore, our probability of drawing a red card would be 26 out of 52 times or if we reduced it to lowest terms, it would be 1 out of 2 times we would get a red card.

Activity 1: Beans

<u>Materials</u> needed: 100 beans

<u>Directions:</u> *The student takes a handful of beans and places them in a container. The student than figures out what fraction of the beans s / he has taken, the ratio of beans taken, and a decimal fraction for the number of beans taken.*

Activity 2: Chain Links

<u>Materials needed:</u> chain links

<u>Directions:</u> sort the chain links according to color.

What fraction of the links is red, yellow, and blue, green?

What is the ratio of red links to total links?

Find other ratios to compare the different colors of the chain links. Roll a die several times and collect that number of a color. Another roll will give a different color. Continue until you have four colors to compare. Write ratios for the different colors.

Rational numbers should be introduced at the Kindergarten level and developed sequentially from that point. Our students have difficulty with fractions; thus they will have trouble with higher level mathematics.

Take time to teach the concept, and build on the student's understanding. Fractions may mean the difference between success and failure in mathematics.

Appendix A

Literature and Math

Note: The literature books listed have been used by students in presenting lessons for the Mathematics Methods classes

Literature Books for Teaching Math

Author	Title	Concept
Abbott, A. Edwin	Flatland	Geometry
Acres, Suzanne	What Comes in Two's, Three's an Four's	Counting, Patterns
Aida, Arlene	Sheep, Sheep, Sheep, Help me Fall Asleep	Attributes
Allen, Pamela	Who Sank the Boat?	Readiness, fractions, basic facts
Anno, Massaichire & Mitsumasa	Anno's Mysterious Multiplication Jar	Multiplication
Anno, Mitsumasa	Anno's Counting Book	Readiness
Balian, Loma	Leprechauns Never Lie	St. Patrick's Day
Bang, Molly	The Paper Crane	Origami cranes, geometry, polygons
Barton, Byron	Dinosaurs, Dinosaurs	Attributes
Baylor, Byrd	Everybody Needs a Rock	Sorting, Classifying, basic facts
Benjamin, Alan	Busy Bunnies	Easter
Bernstein, Stan & Jan	Ready, Get Set, Go!	Generic
Bernstein, Stan & Jan	Trouble with Money	Money
Bernstein, Stan & Jan	Trick or Treat	Attributes
Bethel, Jean	Barney's Beagle	Generic
Birch, David	The Kings Chessboard	Upper Elem.
Bourgeois P. & Clark B	Franklin Fibs	Addition, subtraction, place value
Branley, Franklin	The Planets of our Solar System	Proportion, geometry, measurement
Brenner, Barbara	Mr. Tall and Mr. Small	Readiness, measurement
Brenner, Barbara	Three Little Pigs	Basic Facts
Brenner, Barbara	Too Many Mice	Basic Facts
Brett, Jan	Annie and the Wild Animals	Generic
Brett, Jan	The First Dog	Generic
Brown, Marc	Witches Four	
Brown, Margaret Wise	Seven Little Postmen	Basic Facts
Brown, Margaret Wise	The Runaway Bunny	Basic facts
Burton, Virginia Lee	Mike Mulligan and His Steam Shovel	Geometry
Calder, Lyn	The Little Red Hen	Sorting, graphing, menu math
Calmenson, Stephanie	Dinner at the Panda Palace	
Calmenson, Stephanie	Hopscotch, the Tiny Bunny	Generic
Carle, Eric	Grouchy Ladybug	Time, basic facts
Carle, Eric	Rooster's Off to See the World	Counting
Carle, Eric	Have You Seen My Cat?	Problem Solving
Carle, Eric	The Very Busy Spider	Geometry, measurement, multiplication
Carle, Eric	The Very Hungry Caterpillar	Attributes
Carrick, Carol	Patrick's Dinosaur	Attributes
Charles, Verokika Martenov	The Crane Girl	Geometry, basic facts
Cherry, Lynne	The Great Kapok Tree	Generic
Christlow, Eileen	Five Little Monkeys Jumping on the Bed	Counting
Christlow, Eileen	Five Little Monkeys Sitting in a Tree	Attributes
Clark, Emma C.	The Bouncing Dinosaur	Readiness
Cleary, Beverly	The Growing Up Feet	
Clement, Frank	Counting on Frank	Estimation
Coeer, Eleanor	The Josefina Story Quilt	Patterns, geometry
Cook, Scott	The Gingerbread Boy	Basic facts

Literature Books for Teaching Math

Cooper, Helen	The Bear Under the Stairs	Counting, Addition, Subtraction
Cosgrove, Stephen	Leo the Lop	Attributes
Crews, Donald	Ten Black Dots	Counting, basic facts, creative drawing
Cruicksank, Kathy	The Baby Book	Attributes
Cristaldi, Kathryn	Even Steven and Odd Todd	Even and Odd numbers
Daniel, Frank	The Shape Circus	Geometry
Dee, Ruby	Two Ways to Count to Ten	Readiness
Demi	One Grain of Rice	Problem Solving
DePaola, Tom	The Cloud Book	Attributes, patterns, science
DePaola, Tom	The Popcorn Book	Basic Facts, Problem Solving
DePaola, Tom	Strega Nona	Multiplication
DeLuise, Dom	Charlie the Caterpillar	Counting, basic facts, science
Dodds, Dayle Ann	Wheel Away	Readiness, patterns
Donivee, Martin Laird	The Three Hawaiian Pigs and the Magic Shark	Basic facts
Dr. Suess	The Sneetches	Attributes
Dr. Suess	The Cat in the Hat	Basic Facts
Dr. Suess	The 550 Hats of Bartholomew Cubbins	Estimation, patterns, problem solving
Eastman, P.D.	Go Dog Go	Attributes, generic
Ehlert, Lois	Fish Eyes	Grouping, graphs, story problems
Ehlert, Lois	Color Zoo	Shapes, readiness
Enderle, Judith	Six Creepy Sheep	Readiness
Ernst, Lisa and Lee	The Tangram Magician	Geometry, patterns
Friedman, Aileen	The Kings Commissioners	Place Value
Finger, Charles J	The Tale of Lazy People	Multiplication
Gag, Wanda	Millions of Cats	Place value, classification
Golenbock, Peter	Teammates	Math baseball, percentages, averages
Grifalconi, Ann	The Village of Round and Square Houses	Geometry
Guiberson, Brenda	Cactus Hotel	Generic
Gwynne, Fred	Pondlarker	Basic facts
Hague, Kathleen	Numbears	Number recognition
Hamm, Diane	How Many Feet in the Bed?	Measurement, basic facts
Haskins, Jim	Count Your Way through Israel	Counting
Haskins, Jim	Count Your Way through Japan	Counting
Hayes, Sarah	Nine Ducks Nine	Readiness
Heller, Ruth	Chickens Aren't the Only Ones	Generic
Heller, Ruth	A Cache of Jewels and Other Collective Nouns	Grouping, estimation, number relations
Henrod, Tracey	The Puppy Who Needed a Friend	Attributes
Henry, Marguerite	Misty of Chincoteague	Fractions, Addition
Holtzman, Caren	A Quarter for the Tooth Fairy	Money
Hooker, Yvonne	One Green Frog	Basic facts, logic
Hosen, Michael	We Are Going on a Bear Hunt	Measurement
Hulme, Joy	Sea Squares	Multiplication
Hulme, Joy	Counting by Kangaroos	Readiness
Hutchins, Pat	Rosie's Walk	Readiness
Hutchins, Pat	Clocks and More Clocks	Patterns, problem solving, basic facts, time

Literature Books for Teaching Math

Author	Title	Math Topics
Hutchins, Pat	The Doorbell Rang	Patterns, multiplication, division, fractions
Jeunesse G. & DeBourgoin	The Ladybug and Other Insects	Grouping, basic facts, sorting and classifying
Johnston T. & dePaola	The Quilt Story	Patterns, Geometry
Jonas, Ann	Round Trip	Commutative Property
Kasza, Kieko	The Wolf's Chicken Stew	Place value (100 day)
Keats, Ezra	Whistle for Willie	Attributes
Kent, Jock	The Twelve Days of Christmas	Graphing, logic, multiplication, addition
Khdir, Nash	Little Ghost	Basic facts
Kirk, David	The Spider's Tea Party	Graphing, basic facts, 1-1 correspondence
Kitamuis, Satashi	When Sheep Cannot Sleep	Counting, basic facts
Kraus, Robert	The Littlest Rabbit	Readiness
Kraus, Robert	Herman the Helper	Attributes
Kroll, Steven	The Big Bunny and the Easter Egg	Readiness, basic facts
Laird, Donivee Martin	The Three Little Hawaiian Pigs and the Magic Shark	Basic facts, sequencing
Leaf, Munro	Ferdinand	Attributes, basic facts
Leodhas, Sorche Nic	Always Room for One More	Readiness
Lester, Helen	Tacky the Penguin	Attributes, basic facts
Lester, Helen	Three Cheers for Tacky	Basic facts
Lines, Debbie	Rabbits Night Out	Basic facts
Lionni, Leo	Fish is Fish	Attributes, basic facts
Lionni, Leo	Frederick	Generic
Lionni, Leo	Alexander and the Wind Up Mouse	Basic Facts
Lionni, Leo	Tillie and the Wall	Basic Facts
Lionni, Leo	Swimmy	Basic Facts
Lionni, Leo	Inch by Inch	Measurement
Lobel, Arnold	Frog and Toad all Year Long	Word problems, basic facts, science
Mahy, Margaret	The Seven Chinese Brothers	Ordinals
Mahy, Margaret	17 Kings and 42 Elephants	Multiplication and Division
Mark, David	The Sheep of the Lai Bagh	Measurement
Mathews, Louise	Gator Pie	Fractions
McKee, David	Elmer	Patterning
Milhous, Katherine	The Egg Tree	Generic
Moroney, Lynn	Baby Rattlesnake	Basic facts
Mathews, Louise	Bunches and Bunches of Bunnies	Multiplication
McCloskey, Robert	Blueberries for Sal	Generic
McCloskey, Robert	Make Way for Ducklings	Patterns, counting, basic facts
McGrath, Barbara	The M & M Counting Book	Readiness
McKissack, Patricia	A Million Fish…More or Less	Place value, estimation
Medearis, Angela	Dancing with Indians	Attributes, measurement, graphing, counting
Medearus, Shelf	Picking Peas for a Penny	Generic
Moore, Igna	Six Dinner Sid	Multiplication, prime and composite numbers
Morgan, Allen	Sadie and the Snowman	Attributes, basic facts
Mosel, Arlene	Tikki Tikki Tembo	Generic
Most, Bernard	The Littlest Dinosaur	Measurement, patterns, basic facts
Munsch, Robert	The Boy in the Drawer	Measurement

Literature Books for Teaching Math

Myers, Bernice	The Millionth Egg	Addition, subtraction, place value, estimation, problem solving
Myers, Tim	Let's Call Him Lau-wiliwili-humuhumu	Basic Facts
Myller, Rolf	How Big is a Foot	Measurement, geometry
Nolan, Dennis	Dinosaur Friends	Attributes, sorting, classifying
Numeroff, Laura	If You Give a Moose a Muffin	Measurement, patterning
Numeroff, Laura	If you Give a Mouse a Cookie	Attributes, basic facts, measurement
O"Keefe, Susan Hughes	One Hungry Monster	Counting, fractions
Pallotta, Jerry	The Yucky Reptile Alphabet Book	Attributes, basic facts, measurement
Parind"Aulaire	Don't Count Your Chickens	Counting, basic facts
Payne, Emmy	Katy No-Pocket	Problem Solving
Pfister, Marcus	The Rainbow Fish	Logic, basic facts
Pfister, Marcus	Penquin Pete	Basic facts
Pinczes, Elinor	A Remainder of One	Division
Pinczes, Elinor	One Hundred Hungry Ants	Place Value
Poirier, Thelma	The Bead Pot	Multidigit operations, art, patterning
Pryon, Ainslie	The Baby Blue Cat and the Dirty Dog Brother	Basic facts
Quesada, Esther	Hummingbirds: Jewels in the Sky	Generic
Quindlern, Anna	The Tree That Came to Stay	Generic
Raffi	Baby Beluga	Attributes
Rees, Mary	Ten in a Bed	Subtraction
Ringgold, Faith	Tar Beach	Tesselations, patterns, attributes
Rocard-Morgan, Ann	Kouk and the Ice Bear	Attributes
Rotner & Kreisler	Ocean Day	Attributes, sorting, classifying, basic facts
Ryder, Joanne	The Snail's Spell	Basic facts
Schwartz, David M.	How Much is a Million	Place Value
Shute, Linda	Clever Tom and the Leprechaun	Generic
Scieszka, Jon	The True Story of the Three Little Pigs	Measurement
Sendak, Maurice	Chicken Soup with Rice	Estimation
Sendak, Maurice	Where the Wild Things Are	Generic
Silverstein, Shel	Eighteen Flavors	Graphing, attributes
Silverstein, Shel	The Giving Tree	Attributes, groupings, basic facts
Slobadkins, Esphyr	Caps for Sale	Graphing, counting, patterning, money, basic facts
Slobodkin F. & L.	Too Many Mittens	Counting, patterning, basic facts
Spier, Peter	People	Attributes, estimation
Spier, Peter	The Handmade Alphabet (sign language)	Attributes
Szekers, Cyndy	Moving Day	Attributes
Tafuri, Nancy	Have You Seen My Duckling?	Attributes
Tazewell, Charles	The Littlest Angel	Geometry
Thaker, Mike	The Teacher from the Black Lagoon	Fractions, estimating, logic, attributes
Tompert, Ann	Grandfather Tangs Story	Geometry, patterns
Van Allsburg, Chris	The Polar Express	Attributes, number recognition, basic facts, grouping

Literature Books for Teaching Math

Author	Title	Topic
Viorat, Judith	Alexander Who Used to be Rich Last Sunday	Money
Walt Disney	101 Dalmations	Counting
Walsh, Ellen	Mouse Count	Patterning, basic facts
Weston, Martha	Bea's Four Bears	Basic facts
Wilhelm, Hans	Bunny Trouble	Generic
Williams, Vera	A Chair for My Mother	Counting, basic facts, fractions, money
Williams, Vera	A Chair for My Mother (Un Sillio'n Para Mama')	Attributes
Wood, Don and Audrey	The Little Mouse, The Red Ripe Strawberry and the Hungry Bear	Fractions
Wood, Don and Audrey	The Napping House	Sequencing, attributes, patterns
Wood, Audrey	Little Penquin's Tale	Basic facts
Woodman, June	The Forgetful Spider	Counting, basic facts
Words: J. Kennedy	The Teddy Bears Picnic	Ordinals, counting, basic facts
Young, Ed	Seven Blind Mice	Ordinals, counting, basic facts

Appendix B

GEOBLOCK ACTIVITIES

Geoblock Activities

Geoblocks (a commercial manipulative) are a multicultural manipulative where the language can be changed to the language of need. Geoblocks are used for many activities, and at all grade levels. The Kindergarten student usually begins with building with the blocks and interacting with the various shapes. The seventh and eighth grade student can use them to determine volume, surface area, etc. They can be used to teach geometry, measurement, basic facts, patterning, etc.

The steps which must be identified in a lesson are:

I. Vocabulary – What words do the students need to know? What steps will you take to make sure the ESL students understand the vocabulary?
II. Motivation – How will you get the students interested in the activity?
III. Objective – What do you want the students to do?
IV. Skills needed to learn this lesson - What do the students need to know in order to complete this lesson? What do EL students need to know in order to learn this lesson?
V. Materials – What materials are needed to complete this lesson? (Organize the materials prior to the start of the lesson)
VI. Procedures – What do you want the students to do during the lesson? Will writing be required in this lesson? What study skills are used during this lesson? What will you do with the student whose language is other than English?
VII. Practice – What practice activities will you give the students to help assure understanding of the concept?
VIII. Assessment – How will you assess the activity?
IX. Closure – How will you bring closure to your lesson?

One of the most important factors to consider is "How much can my students complete in one class period?" Be sure you have enough to keep them busy, but not so much as to frustrate them. Lessons can be spread out over several days – remember we need repetition to learn concepts.

Geoblocks are wooden blocks which come in many shapes and sizes. They can be used for many activities. At the onset, geoblocks were part of the Kindergarten program, and were used for activities such as the block center. Some teachers did not like them because they were noisy, and if the correct procedures had not been established, they could cause some problems.

Note: The activities have been adapted from the book Geoblocks and Geojackets from Activity Resource Co. Inc. The geojackets patterns are part of this book.

The activities begin at the primary level and then proceed to the upper grades. If a student has not had experience with the geoblocks at the primary level, the activities should be adapted to allow them the experience to discover the attributes of the geoblocks.

Geoblock Activity

Materials needed: geoblocks file folder, plain paper
Objectives: Students will explore the use of geoblocks and recognize their attributes through play.
Vocabulary: square, rectangle, triangle, face, vertex, edge, area

There are several steps that need to be considered with each activity:
1. ***The mathematical standards which are met by the activity***
2. ***The grade level appropriateness***
3. ***The language needed for second language learners***

Activity 1: Play
Students will play with the geoblocks and build towers, cities, etc.

Students will sort the blocks according to an attribute – square corners, no corners, 4 sides, 3 sides, etc.

Students will find geometric objects in the classroom identify the object and place a post-it note on the object.

Activity 2: Find the Geoblock
Students will select geoblocks to meet the following descriptions:
- A face with four sides
- A face with three sides
- The block with the largest face
- The block with the smallest face
- The vertex of a block
- A block with 4 vertexes
- A block with 3 vertexes
- The largest vertex
- The smallest vertex
- The smallest edge
- The largest edge

Activity 3: Game: Behind the Wall
Two students sit opposite each other with a file folder between them to set up a barrier. Each student should take a set of 5 identical blocks. Players take turns being the building designer and direction follower. Determine who will be the first building designer.

The building designer will use his/her 5 bocks and design a building. Once the building is complete, the designer will give directions to his/her partner to build the building. Each block is described and the position where it must be placed. The partner must listen carefully to the directions and replicate the building. Once the directions are completed, the barrier is removed and the buildings are compared.

Repeat this activity so the second person can be the designer

Activity 4: Game: Behind your back

Each student will need 5 identical blocks. Arrange one set on a table. Keep the other set out of sight. Ask a student to stand with his/her hands behind their back. Place a block in their hands and ask them to describe the block. The person with the block then point to the block which matches the one in the hand.

Activity 5: Art with Block Faces

Trace the faces of the blocks on plain paper. Be sure to trace each face. Cut the faces out and use them to make an art picture. You may use crayons or pencils to add features to your art design.

Trace the faces of the blocks on graph paper. Be sure to trace each face. Cut the faces out. Count the number of squares on each face. Sort the squares according to the size of the faces. Introduce the word **area** at this time.

Activity 6: Geoblock Jackets

Materials needed: Geoblocks, geoblock jackets, colored paper, graph paper, and scissors

Objectives: Students will match geoblock jackets to geoblocks.

Vocabulary: area, surface area

Procedures: Give the students the geoblock jackets and ask them to estimate which block will go with which jacket. After they have estimated the blocks and jackets, fold the jackets and place them around the block.

Sort the geoblocks and only use those which have at least one square face (do not use the triangular faces at this time). Line the blocks up according to the size of one of the faces.

Trace the face on graph paper. Count the squares to determine the area of the face.

Activity 7: Geoblock Maps
Materials needed: geoblocks, geoblock maps,

Vocabulary: geomap, surface

Procedures: Begin with the geoblock map which states you are to use 1 geoblock to cover the surface. Proceed to the map which requires 5 blocks to cover the surface.

Activity 8: Game: The Ant Trip Brain Drain
Materials needed: geoblock jacket for cube, pencil, geoblock to match the jacket, paper (page 48 of the activity book called Geoblocks and Geojackets) which gives the directions. *(Geoblock and Geojackets)*

Vocabulary:

Procedures: Follow the directions given on the worksheet

Activity 9: Symmetry
Materials needed: geoblock jackets

Vocabulary: symmetry, symmetrical

Procedures: Sort the geoblock jackets according to those which are symmetrical and those which are not.

Question: How can you prove that the pieces are symmetrical? Do any of the jackets have more than one line of symmetry?

Activity 10: Match the shapes
Materials needed: geoblocks, pencil

Vocabulary: face, area, centimeter, formula

Procedures: Trace around some of the faces, using graph paper, for the geoblocks. The student tried to match the geoblock to the face that has been traced. Use the centimeter cubes and find the area of each shape. Record your findings. Determine the area of the triangle by using the measurements of the rectangles.

Question: What is the formula for finding the area of a square, rectangle and triangle? How could you teach area without first giving the students the formula?

Activity 11

Materials needed: same as Activity 10

Vocabulary:

Procedures: Use the same faces as in Activity 10, but this time determine the perimeter.

Question: Is there any relationship between area and perimeter. Determine the area of each shape.

Activity 12: Match the Jackets

Materials needed: geoblock jackets, geoblocks which have square or rectangular faces (do not use any blocks that have a triangular face)

Vocabulary: square, rectangular, surface area, area, triangular

> ***Procedures:*** Match the geoblock jackets to the geoblocks. Determine the area of each face of the geoblock.
>
> ***Question:*** What is the total surface area of the geoblock? Can you give the formula for finding the surface area of a shape?
>
> ***Question:*** How would you find the surface area of a geoblock with a triangular face?

Activity 13: Make your own geojacket

Materials needed: geoblocks, geoblock jackets, plain paper, centimeter squares

Vocabulary: volume, similar, congruent, symmetry, triangle, square, quadrilateral, rectangle, hexagon, pentagon

Procedures: Take a plain piece of paper and make a jacket for a geoblock of your choice. Find the volume of one of your geoblocks. (Put the jacket together and fill it with the centimeter cubes.)

Find two blocks which are similar.

Find two blocks which are congruent

Give a definition for the following:

> Symmetry
> Triangle
> Square
> Quadrilateral
> Rectangle
> Hexagon
> Pentagon

The activities have taken the students from the play stage to the developmental stage of being able to determine their own formulas. They have interacted with the manipulatives, and have abstracted the attributes of each of the models.

A good culminating activity which can be used at the upper grade level is the State of California replacement unit called Polyhedraville. This unit follows the activites of the Geoblocks, and enables the students to build their own cities, and apply what they have learned.

Appendix C

APPLE ATTRIBUTE TREE

AND

ATTRIBUTE TREE

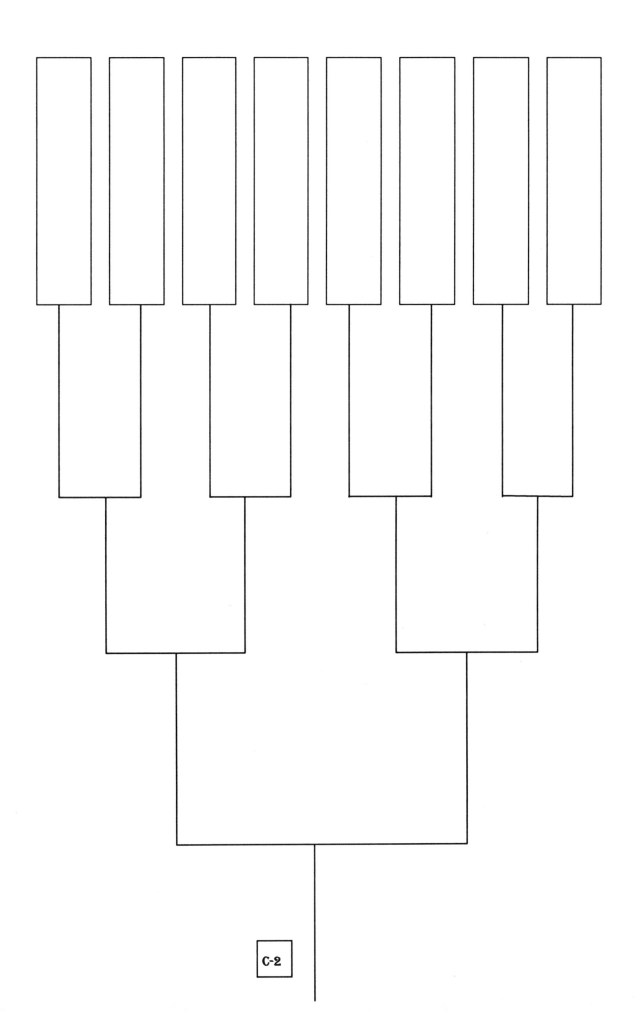

Apple

C-3

Color	Size	Leaf/No Leaf	Worm/No Worm

- Apple
  - Large Apple
    - Large Red Apple
      - Large Red Apple with Leaf
        - Large Red Apple with Leaf with Worm
        - Large Red Apple with Leaf No Worm
      - Large Red Apple with No Leaf
        - Large Red Apple with No Leaf Worm
        - Large Red Apple with no Leaf no Worm
    - Large Green Apple
      - Large Green Apple with Leaf
        - Large Green Apple with Leaf Worm
        - Large Green Apple with Leaf no Worm
      - Large Green Apple with No Leaf
        - Large Green Apple with no Leaf no Worm
        - Large Green Apple with no Leaf no Worm
  - Small Apple
    - Small Red Apple
      - Small Red Apple with Leaf
        - Small Red Apple with Leaf with Worm
        - Small Red Apple with Leaf with no Worm
      - Small Red Apple with No Leaf
        - Small Red Apple with no Leaf Worm
        - Small Red Apple with no Leaf no Worm
    - Small Green Apple
      - Small Green Apple with Leaf
        - Small Green Apple with Leaf Worm
        - Small Green Apple with Leaf no Worm
      - Small Green Apple with No Leaf
        - Small Green Apple with no Leaf no Worm
        - Small Green Apple with no Leaf no Worm

Appendix D

Attribute Activities

TEDDY BEAR MATH

Teddy Bear! Teddy Bear! Can you help me do my math? Yes, oh, yes I can help you do your math. Just tell me what math you want to do, and I will find a way to help you.

Manipulatives: Students' teddy bears
teddy bear attribute models
die cut teddy bears
Pom-pom teddy bears
teddy bear counters

Math Concepts: Patterning
Attributes
Logic
Basic Facts

Teddy bears bring fun and excitement to mathematics learning. Children can interact with the teddy bears, solve problems, and develop logical thinking skills at the same time. Teddy bears can help the children develop a positive attitude toward learning mathematics and empower them to be able to communicate mathematically.

The teddy bears will lead the children into various activities that will help them develop thinking skills, write about mathematics, and develop a mathematical **Vocabulary**.

The NCTM Standards states that children need to be active learners of mathematics and have a curriculum that is developmentally appropriate. They need to use manipulatives and interact with these manipulatives.

TEDDY BEAR WEEK
One week prior to having Teddy Bear week at school, send a letter home to the parents asking the children to bring their teddy bear to school (if students do not have a teddy bear, find a way to share teddy bears).

Introductory Activity: Read a story such as Corduroy by Don Freeman

---
Activity 1: Teddy Bear Comparison
Math Concept: Likenesses and differences
Materials: teddy bears
Procedure: Each student will introduce their teddy bear to the class. They will describe their own teddy bear, listing as many attributes as they can.

Activity 2: My Teddy Bear is Like Yours!
Math Concept: Attributes
Materials: teddy bears
Procedure: Students find a partner that has a teddy bear similar to theirs. They verbalize the likenesses and differences. Since there are so many possibilities with this activity, this activity could go on for several days.

Extension: Place several large circles (i.e. rope or hula hoop) on the floor. Place words (i.e. large and small or colors, etc) in the circles that describes the teddy bears. The students can determine if their teddy bear will fit in one of the circles.

Activity 3: Graph the Teddy Bears
Math Concept: Graphing
Materials: Large sheets of paper, crayons, pencils, teddy bears, scissors
Procedure: The students lay their teddy bears on a sheet of paper, and trace around them. They color their teddy bear to match the real one, and cut it out. After all of the teddy bears have been cut out, they are placed on the wall according to height.
Extension: Discuss the differences in height. Compare two different teddy bears, Ask the students to write a sentence comparing several teddy bears.

Activity 4: How Tall is my Teddy Bear?
Math Concept: Measurement and graphing
Materials: Teddy Bear cut outs (those used in the previous activity), unifex cubes (or any other measurement tool you have available that all students could use).
Procedure: The students use a measurement tool such as unifex cubes and determine the height of their teddy bear. They record this measurement and give it to the teacher. As a Total Group Project, each of the measurement is read and recorded on a graph.
The teacher than asks the children if more than one teddy bear is the same height. Once they discover the height common to the most teddy bears, the term mode can be introduced.
The teacher can also help the students discover the mean height of the teddy bears.

Activity 5: How Much Does My Teddy Bear Weigh?
Math Concept: Measurement
Materials: Pan balance, weights, teddy bears, pencil, paper
Procedure: Students place their teddy bears on one side of the pan balance, and weights on the other side. They count the number of weights to determine the weight of their teddy bear.
Extension: Students estimate the weight of each teddy bear. After all teddy bears are weighed, each student checks their own estimate.
A class graph off each teddy bear's weight can be made.

ACTIVITIES USING TEDDY BEAR COOKIES

Activity 1: The Teddy Bears are different!
Math Concept: Sorting and Classifying
Materials: several different flavors of teddy bear cookies
Procedure: Each student is given several flavors of teddy bears. They sort and classify their teddy bears according to flavor.

Activity 2: Taste Test
Math Concept: graphing
Materials: teddy bear cookies, class graph
Procedure: The students taste the cookies in their baggy. They must decide which one is their favorite cookie. They tell the teacher, and the teacher records it on the class graph. After each one has given their preference, the teachers helps the students determine the favorite flavor, etc.

ACTIVITIES USING TEDDY BEAR ATTRIBUTE MODELS

After the children have a variety of experiences with sorting and classifying themselves and objects in the classroom, they are ready to begin to use the teddy bear models. The following activities are designed to use with either a large set of teddy bears on a 9 x 12 card or a set of small teddybears.

The teddy bears used in this activity had four general characteristics or attributes: **size** (large or small); **color** (brown or black); **color of tip of no**se (red or black) and **eyes** (open or closed). A model with four characteristics will make a 16 piece attribute model: 2 sizes x 2 colors x 2 colors of noses x 2 types of eyes equals 16 pieces.

The following activities are designed to be used with either small groups or a total class. A large model of the teddy bears (placed on 9 x 12 oaktag – teddy bears can be made with different sized circles) should be constructed and used for introductory activities while several regular sets of the models will be used for group work. When the models have been completed and prior to using them in the classroom, place them in a learning center for a period of free play.

Activity 1: What are my characteristics (attributes)?
Math Concept: Attributes
Vocabulary: names of the attributes (large, small, brown, black, red nose, black nose, open eyes, closed eyes)
Materials: attributes models
Procedure: Give one of the 9 x 12 teddy bear attributes to each of 16 children. Each child will describe their teddy bear and the rest of the children will listen. If an attribute is missed, the other children in the class will add the attribute. Discuss the attributes of all of the teddy bears.

Activity 2: Teddy Bear Meeting
Math Concept: Attributes
Vocabulary: names of attribute, meeting
Materials: 9 x 12 models of teddy bears
Procedure: The teddy bears will now be grouped according to some characteristics. Example: The teacher will group the small teddy bears to have a teddy bear meeting. The small teddy bears will go to the meeting. The rest of the class will decide if the right teddy bears are at the meeting. The teacher will change the characteristics of the teddy bears that will go to the meeting. After several meetings, the teacher will add another characteristic, i.e. I want to meet with the small brown teddy bears. This can continue until the teacher has added several characteristics to the teddy bear meeting.

Activity 3: You Belong Together
Math Concept: sorting and classification
Vocabulary: belong together, sort, classify
Materials: a set of 16 teddy bears for each group, word cards that label the characteristics (large, small, brown, black, open eyes, closed eyes, red nose, black nose)
Procedure: Ask the children to sort the teddy bears in some manner. Tell them to place the word on the top of their desk that tells how they sorted their teddy bears; i.e. if they sorted by size, they will display the words large and small above their groups. After all of the groups have grouped their teddy bears in some manner, ask them to tell the class how they sorted their teddy bears. Ask them if they can sort their teddy bears in a different manner.

Activity 4: I'm in a circle
Math Concept: Venn Diagrams
Vocabulary: attribute words, circle
Materials: one set of teddy bears for each group, two large circles (these can be made of rope or plastic ones can be purchased) words describing the attributes.
Procedure... Give each group two circles and the attribute words. Ask them to find the word (or rhebus picture) for large and small and place one word in each of the circles. Next ask the children to sort the teddy bears according to the words in the rope. Once they have successfully completed this, ask them to place two different words (be sure to use only the words that go together such as the size - large or small; color brown or black; eyes - open or closed; and nose - red or black) in the circles.
Extension: Place the teddy bear cards, circles and word cards in a learning center for further exploration. The cards will only feature one major characteristic.

Activity 5: Teddy Bear Train
Math Concept: Likenesses and differences
Vocabulary: one different
Materials: large model of teddy bears on 9 x 12 oaktag, 16 chairs in the front of the room
Procedure: One teddy bear is randomly chosen to sit in the middle chair. The rules for the teddy bear train are any teddy bear that sits next to someone else can only have one attribute different i.e. if the Large black teddy bear with a red nose and open eyes sits in the middle chair, the teddy bear who sits on either side must have only one attribute different.
Example: a small black teddy bear with a red nose and open eyes could sit in a chair (one different - size); or a large black teddy bear with a red nose and closed eyes could sit in a chair (one different - eyes); or a large brown teddy bear with a red nose and open eyes could sit in a chair (one different - color).
As the children join the teddy bear train they must verbalize why they should be able to sit in the chair.

Activity 6: I'm Different by One Attribute
Math Concept: Likeness and Differences
Vocabulary: same as activity 5
Materials: one set of teddy bears for each group of children

Procedure: Give each group of children one set of teddy bears. The groups will randomly choose a teddy bear and place in the center of the group. They will spread the rest of the teddy bears out for everyone to see. The game will follow the same format as Activity 5, except this time the small group will make the one different train. As each child places a teddy bear on the one different train, they must verbalize the difference. The other children will need to be good listeners as they will determine if the teddy bear can be added at that particular place.

Extensions: After the children become proficient with a one different train, a two or three different train can be added.

Activity 7: Teddy Bear Riddles
Math Concept: Language, Critical Thinking
Vocabulary: attribute words, any other words the children do not know
Materials: large model of teddy bears on 9 x 12 oaktag, riddles the teacher has written (either on large chart paper or on the overhead projector)
Procedure: Place all of the teddy bears on the chalk tray (or on a wall) in front of the room so all of the children can see them. The teacher will ask the children to pick their favorite teddy bear and not to tell anyone. The teacher will then begin to describe a teddy bear by displaying one sentence at a time.

I am a small teddy bear

If the teddy bear chosen by the child is not a small teddy bear, they will need to choose a small teddy bear (the teacher might want to remove all of the large teddy bears at this time).

Additional sentences are displayed (and teddy bears removed that do not fit the description) until the children have decided which of the teddy bears the teacher has described.
My eyes are open
I have a black nose
I am black

(Ans: Small black teddy bear with open eyes and a black nose.)

The language of the card can gradually be made more difficult, i.e.
I am not a large teddy bear
My eyes are not open
I have a red nose but I am not brown
I am not large and my eyes are not open
I do not have a red nose or closed eyes

Rhebus pictures can be used for young children. For example a small black dot could represent a black nose; or an eye that matches the open a closed eyes could be used.

Activity 8: I'm Writing about My Favorite Teddy Bear
Math Concept: Language
Vocabulary: words the children will need to write about one of the teddy bears.
Materials: One set of large teddy bears, paper and pencil

Procedure: Place a large set of teddy bears on the bulletin board and tell the children they are going to write a story about their favorite teddy bear. The story is like a riddle and will make the rest of the boys and girls guess which teddy bear they are writing about. (Follow the procedures for the writing process: rough draft, conference, revision, and final paper.) Have the children make a rough draft of their story. During conference time, the teacher will help the children with their story and help them correct their spelling. After the children have written their final draft, they will place their story on the bulletin board. Others in the class will read their story and try to determine which of the teddy bears it is describing. If they think they know the teddy bear, they will place the story next to the teddy bear. If the student who wrote the story finds the wrong teddy bear by their story, they will remove the story and place it at the bottom of the bulletin board. Once all of the stories have been matched to the teddy bears, they class will discuss the stories and teddy bears. A teddy bear book could be made for the class to read.

Activity 9: A Teddy bear is lost!
Math Concept: Critical Thinking
Vocabulary: lost
Materials: large set of teddy bears on 9 x 12 oaktag
Procedure: Choose one child to take a teddy bear out of the pile. Ask the child to hide the teddy bear in a large envelope. The child that has chosen the teddy bear to hide, now helps the teacher. The other 15 teddy bears are given to other children. The teacher will tell a story about a little lost teddy bear and ask the rest of the children to tell which teddy bear is lost. A discussion will follow as to how they will discover which one is lost. Let the children suggest ways to make the discovery. As they suggest a teddy bear, if a student as that teddy bear, they must speak up and let the rest of the class know they are not lost. When they have discovered the lost teddy bear, the teddy bear is brought out of the envelope and shown to the rest of the class.

Activity 10: Help Me! I Don't Know Who I Am!
Math Concept: Attributes
Materials: one large set of teddy bears on 9 x 12 card
Procedure: Place a string on the large teddy bear cards and place one on the back of each of 16 children (be sure the children do not see which one you place on their back). Tell the children they must discover which teddy bear they have on their back. The only rule is all questions can only be answered with a "yes" or a "no". Once they discover who they are, they will go to the front of the room and wait for the rest of the teddy bears. When all of the teddy bears have joined the group at the front of the room, each will describe their teddy bear.

Activity 11: Match Me!
Math Concept: Likenesses
Materials: set of teddy bears, cards to describe the teddy bears
Procedure: This is a game for two children. Make a concentration game of the teddy bears and the descriptions. The children will place all of the teddy bears face up and all of the cards describing the teddy bears face down in front of them. They will take turns. Each will turn over a card that describes a teddy bear and match it to the teddy bear. If they make an incorrect match, they must

replace it, and the other child will get a turn. Play continues until they have matched the teddy bears and the descriptions.

Activity 12: I'm Going on a Picnic!
Math Concept: Listening
Materials: set of teddy bear cards, pretend blanket for picnic
Procedure: A pretend picnic is set up in the classroom. Someone begins the picnic and tells the teddy bears, s/he is going on a picnic. That teddy bear will now describe the other teddy bear that is now going on a picnic. Play continues until all of the teddy bears have joined in the picnic.

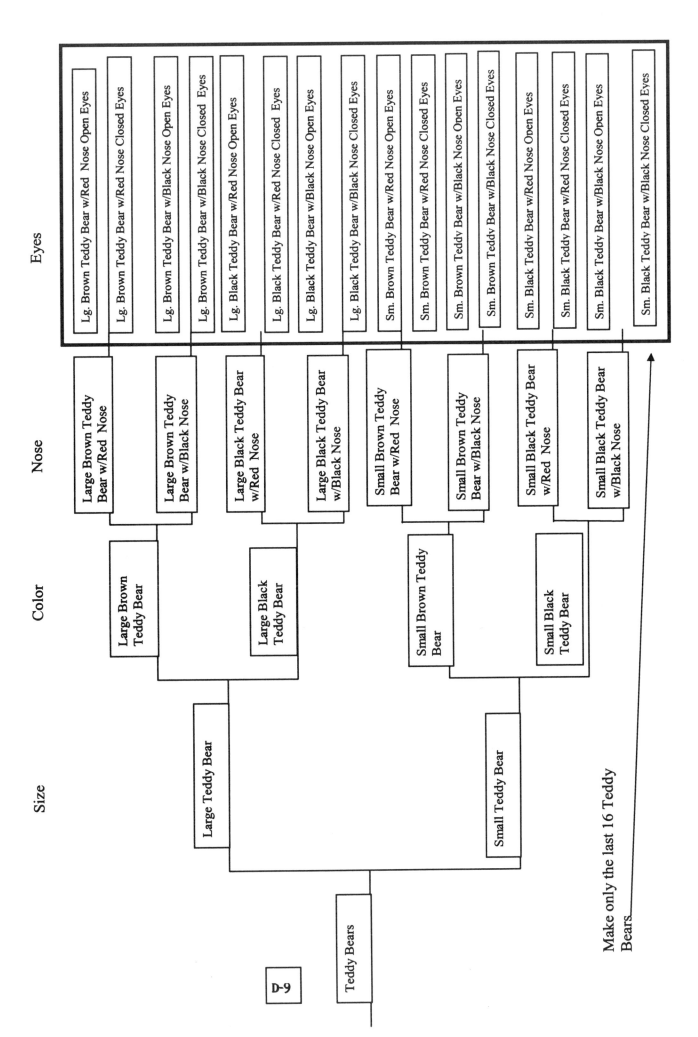

Fish Attributes

Books: Swimmy by Leo Lionni

The book Swimmy by Leo Lionni is a good motivator and can be used to develop many mathematical concepts. Swimmy is a little fish in a big sea. Swimmy looses all of his friends and must find more.

Manipulatives:

Origami Fish
Cracker or cookie Fish
Lima Bean Fish

Directions: One side of the lima bean can be painted orange and the other side yellow - or use the colors of your choice - or spray them all one color. The fins can match the sides of the fish or they can be only one color. The side of the fin that touches the lima bean can be curved to make it fit on to the lima bean.

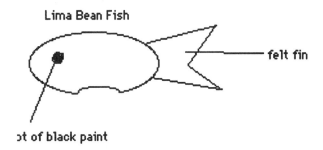

Game Boards

A file folder is used for the game board. The bottom of the file folder represents a scene from the bottom of the sea. The top represents a scene from the top of the sea. (The fold line is a natural division line.) Paints, or cut-outs can be used to make the scene. (This would be a good art project for the students in the class.) Fabric paint is easy to use, but care needs to be taken so the top and the bottom of the folder do not stick together.

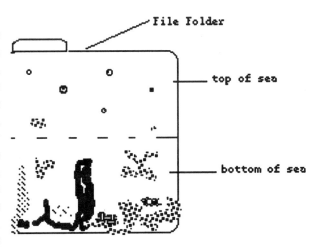

Math Concepts:

Sorting and classifying
Logic
Comparisons

Basic Facts
Place Value
Problem Solving

Suggested activities that have been successfully used with students are:

Activity 1: Swimmy –
   Read and discuss the story; begin vocabulary development

Activity 2: Origami fish
Math Concept: *Sorting, Classifying, Logic*
Materials: *Origami Fish*
Procedures: *One of the first concepts that we want children to have is the ability to sort and classify. Give each group a set of origami fish and let them play with the fish. Ask them to sort them in some manner. Discuss the various ways they were sorted.*
Game: *Play the one difference train game. Place all of the fish on the table and randomly choose one fish. This will be the beginning of the train. Student will play a fish next to the first one (it can be either on the left or right side). The fish must have only one attribute different from the one played. The students verbalize their reasoning for placing a particular fish. Play continues until all of the fish have been played.*

Activity 3: Cracker Fish
Math Concept: *Estimation, graphing*
Materials: *bag of assorted goldfish crackers, square graph paper, crayons*
Procedures: *Each group will have a bag of fish crackers (different kinds).* **Note:** *Some of the fish flavors are the same color. You might want to use only those where there is a distinct difference.*
(1) Estimate the number of gold fish in the bag and record the estimate. (the number of fish in the bag will depend on the age level of the students).
(2) Sort the fish according to kinds.
(3) Place the fish on the square graph paper according to the kind, i.e. all pretzel fish will be grouped together and all cheddar cheese fish will be grouped together. Place one fish on each square. You will be making a concrete graph.
(4) Remove one fish at a time and draw a picture of the fish as you remove it. This becomes your pictorial graph.
(5) Color in the squares to make your bar graph.

Activity 4: *Group those fish into a big fish.*
Math Concept: *Grouping*
Materials: *bag of cracker fish, gameboard*

Procedures: Ask the question: "How many fish will you need to make a large fish?" Have the students record their estimate and then make the large fish on the gameboard. How many large fish can you make from your bag of fish?

Is there any way you can make 3 large fish from your cracker fish?

What is the largest number of large fish you can make from your cracker fish? Remember you have to make a large fish like Swimmy did so the other large fish in the sea will not eat you.

Activity 5: Taste Test

Math Concept: *graphing*

Materials: same as in Activity 3.

Procedures: Write the names of the types of fish on you graph, i.e. pretzel, cheddar cheese, etc. Taste the fish and decide which kind you like the best. Color in the square of the type you like the best.

After all groups have completed this, a large class graph can be made. For older children the term - mode, median and average can be introduced.

Activity 6: Are the Groups the Same or Different

Math Concept: Equal, Not Equal

Materials: Lima bean fish, or origami fish, or cracker fish

Procedures: Tell a story about the fish Monster (You might want to design some sort of a fish monster to go with this story.) The fish monster only likes fish that are in groups that are not equal. If the groups of fish are equal the fish monster will swish its tail and scare all of the fish. Randomly generate two numbers. Place one group of fish on top of the sea (file folder game board) and one group on the bottom of the sea. Discuss whether the fish monster would be happy or swish his tail.

Seatwork: Students randomly generate a number and place that number of fish on the top of the gameboard. They generate a second number and place that number of fish on the bottom of the gameboard. They can then write their sentences to tell if the fish monster is happy or sad (equal groups). They can also be taught the sign for equal (=) and not equal (=). The not equal sign is an equal sign with a line drawn through the middle.

Activity 7: Which group will the shark eat?

Math Concept: Number recognition, Greater Than, Less Than

Materials: lima bean fish, or origami fish, die or digit cards (cards with one digit 0-9 written on a card). gameboard.

Procedures: *Randomly generate two numbers. Place one group of fish on top of the sea (file folder game board) and one group on the bottom of the sea. Tell a story about the shark always wanting to eat the most of everything. Ask the students which school of fish the shark would like to eat.*

<u>Group Work</u>. *Each person in the group will roll a die, or draw a digit card. The number that is rolled or drawn tells how many fish that person must collect. After each person has a turn, they compare the number they have. They can also write a sentence at this time - i.e. John has 5 fish and I have 2 fish. John has more than I do.*

Extension: *The shark swims the ocean in search of fish. The shark is always hungry and makes sure it eats the most of everything. Randomly generate two numbers and build those numbers. Introduce the < > signs and determine which of the groups the shark would want to eat.*

Activity 8: Swimmy Plays Games
Math Concepts: Readiness for Addition and Subtraction.

Materials: *fish, gameboard*

Procedures: *Expand the story of Swimmy.*

Some of Swimmy's friends decided to play games. They wanted to hide from Swimmy and have Swimmy guess how many fish were hiding. They decided it would be best if two groups of fish hide, which would make it harder for Swimmy.

(Students will model the story as it is told). Four little fish went to hide behind a little pebble in the sea. Three other fish hide behind some leaves. Once they were hidden, they called to Swimmy, "Come and find us". Swimmy decided he needed to know how many fish he was going to find, so he called out, "give me some clues so I will know how many fish I need to find. The first group of fish each called out their numbers - one, two, three, four." Swimmy asked if there were four in the first group - the answer was yes. Swimmy decided if he was going to remember the number, that he should write the number in the water - so he wrote a 4 (have the students record a 4). Swimmy asked the second group how many were in their group - they answered one, two, three. Swimmy wrote a 3 in the water. Swimmy now has a 4 and a 3 written in the water. He needs to know how many he has altogether. Can you help Swimmy determine the number of fish that he must find? (Students can add the fish or count the total number of fish and then record their answer.) Swimmy looked and looked and found the four gold fish. He knew that he must still look for more. He finally found the other three gold fish. He decided he better count the gold fish to see if he did have seven. He counted 1, 2, 3, 4, 5, 6, 7. "Hurray" said Swimmy, "I found all of you." "Let's play this game again", said the fish.

Several different activities can be completed by giving different numbers. The number of fish will depend on the level of the students. If you want to use two digit numbers, the origami fish could be used as a representation of the ten and the crackers as the ones, or a cup of ten crackers could represent the ten and the single crackers could be the ones, or if you have lots of origami fish, the large origami could be the ten and the small fish could be the ones.

Extention: Students will randomly generate a number and place that number on the gameboard. They will randomly generate a second number and place that number on the gameboard. They will record both numbers and determine how many they have altogether.

Activity 9: Some fish are missing.

Math Concept: Readiness

Materials: fish and gameboard

Procedures: Tell the story of some fish that are missing.

A school of fish was swimming in the sea and they got very close to some of the big fish. They thought they were in the safe waters, so they did not worry. Five little fish went to tell their mommy they were going out to play. Mommy said it was alright, but warned them of the dangerous big fish in the water. The five little fish happily went off to play. They were not paying attention to where they were playing. Pretty some there were only three of the fish playing. Some were missing! The other little fish tried and tried to figure out how many were missing. Can you help them. Five started to play and now only three were playing. How many are lost. (If they are using the cracker fish, they can eat the number of fish that were missing.)

Extended Activities: Two random numbers are generated and the larger number of fish are placed on the gameboard and the smaller number the fish lost. Students will record their actions in some manner.

Activity 10: We Must be the Same!

Math Concept: *Basic facts - addition and subtraction*

Materials: *same as activity 9 plus the game board*

Procedures: <u>Children work in pairs</u>. Randomly generate two numbers and place one group of fish on the top of the gameboard and one group of fish on the bottom of the gameboard. The two groups must be equal, however, only one group can be changed. If there are 5 on top and 6 on the bottom, you have the option of adding to the 5 or taking away from the six to make the two sets equal. (This activity allows the students the option of either adding or subtracting to make things equal. This also allows for teaching addition and subtraction as inverse operations rather than separate facts.)

Again the random number generators can determine the number on the top and the bottom of the sea. This activity will aid the students in solving problems that deal with the missing addend and the equalization will be a prerequisite skill needed when they begin to do algebra.

Activity 11: A Fish Sentence

Math Concept: *Equations for addition and subtraction*

Materials: *eggs with lima beans (same as fish but without eyes and tails)*

Procedures: *Students will need to count the number of beans in the eggs and then replace them in the egg. One person in the group says add or subtract - that is the operation that will be completed with the beans. The beans are spilled on the table and the number of yellow and orange fish are counted. The students then write a number sentence using the operation stated. For example if the egg has 7 beans and someone in the group calls for subtract. The sentence would begin with the total amount - 7 and then take away the yellow or the orange. Once this sentence is complete, the beans are replaced in the egg and another sentence is generated.*

Activity 12: My Friends!

Math Concept: *Place value*

Materials: *lima bean fish*

Procedures: *All of the fish decided they wanted to play a game, but they could not decide how to make the groups needed for the game. Finally one little fish decided there should be five fish in each group. They talked and talked about how to make groups of five and finally decided the best way to do it was to just count - one, two, three, four and five. Once they got to five they started over. They would like to have you help them make groups of five with the fish that you have. Write a sentence telling how many groups of five you have and how many extra fish. Now another little fish decided they should change the groups. Ask one of your group members to decide on another way to group the fish (Make sure the groups are less than 10). This can be repeated several different times.*

Activity 13: Those Fish Multiply

Math Concept: *Multiplication and Division*

Materials: *lima bean fish*

Procedures: *The fish are going on a vacation and want to stay at the fish hotel. They decided that four fish should stay in each room. They reserved four rooms. How many fish are going on the trip?*

Reverse: There are 16 fish going on a trip. They want to have four fish in a room. How many rooms will they need?

Repeat each of the following using different numbers.

Activity 14: Fish Train

Math Concept: *Multiplication*

Materials: *lima bean fish, graph paper, large fish*

Procedures: *The little fish want to go to the fish playground, however, the sharks in the sea keep eating the little fish. The older fish decided they must design a fish train to take the little fish to the playground. They asked the little fish to help them design a train. Each group had to decide how many fish would be on the train and how they would design the seats so there would be an equal number of rows and seats - the design must form a rectangle - if they decide on a train for 30 fish, they could either have 30 rows of 1 seat in each row; or 6 rows of 5 in each row; or 5 rows of 6 in each row; or 15 rows with 2 in each row; or 2 rows with 15 in each row.*

Design your fish train and decide how the number of rows and the number of seats in each row. Write your rationale for designing the seats in the manner you did.

Activity 15: Divide those fish

Math Concepts: Division

Materials needed: *Egg carton trains or fish trains similar to the one above, fish*

Procedures: *Cut egg cartons into different groups - 0 through 9. Remember the egg cartons must form a rectangle therefore a odd number must be a rectangle such as a group of 3 whereas an even number such as six could be six cups, or 2 groups of three. Three sets of random number generators, the first will tell the number of egg cartons needed (i.e. a 5 is rolled, they choose an egg carton with 5 cups) The next two random number generators will build a two digit number which will be divided evenly between the cups.*

Example: *Roll 1 is a 6. A 6 egg carton will be found. The next two numbers are a 3 and a 7. Either a 37 or a 73 could be made. (You might want to tell them to make the smaller number at the start of the activity. After they have had more experience, they can use the larger number.) The students then take the 37 fish and divide them equally between the 6 cartons. They will find that 6 will go into each cup and there will be 1 left over. Left over become remainders as they begin to do division activities.*

Other attribute models are listed below, with manipulatives which can be made to enhance the mathematical concepts.

Frogs - Origami frogs, frogs made from lima beans. The frogs are made from lima beansprayed green, lentils are used for

- **Pigs** - Pom Pom Pigs, lima bean pigs (pink pom poms for the pigs - large for body, medium for head, two small ones for each leg and a tiny one for the nose. The ears are pink foam, and the tail is a pipe cleaner. The lima bean pigs are sprayed pink with foam for the ears, t-shirt paint for the eyes and nose.
- **Mice** - pistachio half shells sprayed gray (ears are gray foam, eyes are dots of black T-shirt paint, and tails are pieces of yarn)
- **Turtles** - walnut shells sprayed green, felt turtle shapes glued on to the bottom of the walnut shells.
- **Apples** - pom poms in the color of the apples. A dry stick and be used for the stem of the apple, and brown t-shirt paint for the blossom end. Gabanzo beans can also be sprayed red yellow and green to represent apples when larger numbers are needed.
- **Feet** - die cut feed are available, children can draw around their feet and cut them out to use for various measurement activities.
- **Snails** - pasta shells can be dyed using rubbing alcohol and food coloring; pipe cleans can be used for snail bodies
- **Caterpillars** - spiral pasta can be dyed using rubbing alcohol and food coloring; t-shirt paint can be used to make the eyes
- **Butterflies** - pasta bows dyed can be used or small snap clothes pins which have tissue paper, crepe paper, or material folded into the snap - features can be drawn on the butterflies
- **Paper Cranes** - origami cranes can be made by older children.
- **Rattlesnakes rattles** - beads can be used to make the rattle tails of rattlesnakes

Appendix E

Addition Chart
Multiplication Chart

+	0	1	2	3	4	5	6	7	8	9
0	0	1	2	3	4	5	6	7	8	9
1	1	2	3	4	5	6	7	8	9	10
2	2	3	4	5	6	7	8	9	10	11
3	3	4	5	6	7	8	9	10	11	12
4	4	5	6	7	8	9	10	11	12	13
5	5	6	7	8	9	10	11	12	13	14
6	6	7	8	9	10	11	12	13	14	15
7	7	8	9	10	11	12	13	14	15	16
8	8	9	10	11	12	13	14	15	16	17
9	9	10	11	12	13	14	15	16	17	18

Strategies are used at each level starting with 0 plus a number is that number to a number added to 9 is one less than the number with a 1 in front of it, i.e. in 7 + 9—one less than 7 is 6 and place a 1 in front of it and you have 16.

The facts considered hard are the blue, yellow and green facts.

X	0	1	2	3	4	5	6	7	8	9
0	0	0	0	0	0	0	0	0	0	0
1	0	1	2	3	4	5	6	7	8	9
2	0	2	4	6	8	10	12	14	16	18
3	0	3	6	9	12	15	18	21	24	27
4	0	4	8	12	16	20	24	28	32	36
5	0	5	10	15	20	25	30	35	40	45
6	0	6	12	18	24	30	36	42	48	54
7	0	7	14	21	28	35	42	49	56	63
8	0	8	16	24	32	40	48	56	64	72
9	0	9	18	27	36	45	54	63	72	81

Strategies are used at each level starting with 0 times a number is zero. The numbers for 9 x a number must add up to 9. Some remember 7 x 8 as 5, 6, 7, 8 or 56. Discover the patterns for all of multiplication.

Appendix F

Fraction Bars
Decimal Bars
Percentage Bars

1 Whole

1/2

1/3

1/4

1/5

1/6

1/8

1/9

1/10

1/12

1/15

1/16

1/18

1/20

1/24

1.0

.50

.333

.25

.20

.166

.11

.10

1/10

.083

.0666

.0625

.055

.05

.0416

100%

50%

33.3%

25%

20%

16.6%

12.5%

11%

10%

8.3%

6.66%

6.25%

5.5%

5%

4.16%

Appendix G

Historical Checking Strategies

CHECKING FOR LONG ADDITION

$275 =$ $2 + 7 + 5 = 14$ $1 + 4 = 5$

$863 =$ $8 + 6 + 3 = 17$ $1 + 7 = 8$
$\qquad\qquad\qquad\qquad\qquad\qquad 8 + 5 = 13 \quad 1 + 3 = 4$

$429 =$ $4 + 2 + 9 = 15$ $1 + 5 = 6$

$\underline{\ 687} =$ $6 + 8 + 7 = 21$ $2 + 1 = 3 \quad 6 + 3 = 9\text{-}9 = 0$
$\qquad\qquad\qquad\qquad\qquad\qquad 4 + 0 = 4$

2254 $\qquad\qquad 2 + 2 + 5 + 4 = 13\text{—}9 = 4$

The Casting Out Nines may seem to be cumbersome with all of the steps, however, if it helps one student succeed it is worthwhile. Each number is added together until it is a single digit number.

1. $2 + 7 + 5 = 14$ $\quad 1 + 4 = 5$
2. $8 + 6 + 3 = 17$ $\quad 1 + 7 = 8$
3. Add the single digits from the first 2 numbers. $5 + 8 = 13 \quad 1 + 3 = 4$
4. $4 + 2 + 9 = 15$ $\quad 1 + 5 = 6$
5. $6 + 8 + 7 = 21$ $\quad 2 + 1 = 3$
6. Add the single digits for the last 2 numbers. $6 + 3 = 9$.
7. Cast out the 9 which leaves 0. $9 - 9 = 0$
8. Add the digits in the answer $2 + 2 + 5 + 4 = 13 \quad 1 + 3 = 4$
9. The single digit for the numbers being added is 4 and the single digit for the answer is 4, therefore the answer is correct.

Dutch method

	30	6	
20	600	120	720
4	120	24	144
	720	144	864

This method is the next step in using the numeration blocks. Separate the tens and ones.

1. 30 x 20 = 600
2. 20 x 6 = 120
3. 4 x 30 = 120
4. 6 x 4 = 24

Add the numbers going across

5. 600 + 120 = 720
6. 120 + 24 = 144
6. 720 + 144 = 864

Next Add the numbers down

8. 600 + 120 = 720
9. 120 + 24 = 144
10. 720 + 144 = 864.

MULTIPLICATION EXPANDED NOTATION

$$\begin{array}{r} 231 \\ \underline{\times\ 122} \end{array}$$

<u>2</u> X 231	462
20 X 231	4620
100 X 231	<u>23100</u>
	28182

Focusing on the ones, tens and hundreds may help some students successfully understand multidigit multiplication.

1. The factor 122 is divided into 2 ones, 2 tens (20) and 1 hundred (100)
2. 2 x 231 = 462
3. 20 x 231 = 4620
4. 100 x 230 = 23100.
5. Add the 3 answers together and we get 28,182.

CASTING OUT NINES FOR DIVISION

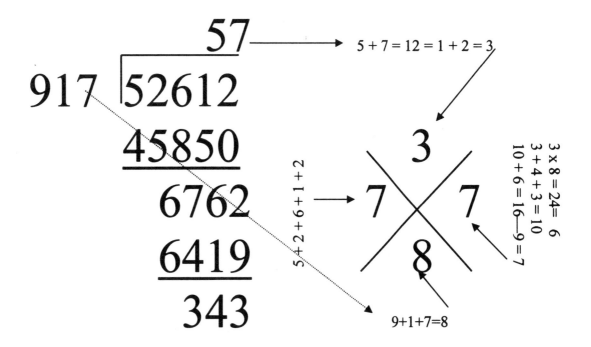

Casting out nine for division is more complex.
1. Take the factor in the answer—57 and add the two digits—12. Take 9 away (12-9) =3. Place the 3 on top of the x.
2. Take the second factor—917 and add the digits. Since there is a 9, cross it out and add the 1+ 7. Place the 8 at the bottom of the x.
3. Multiply the top number 3 x the bottom number 8. (3x8=24). Add the 2 digits which gives us 6.
4. Add the digits from the remainder 3 + 4 + 3 which gives us 10. Add 10 + 6 =16. To get a one digit number add 1 + 6 = 7. Place the 7 on the right side of the x
5. Add the digits of the product. 5 + 2 + 6 + 1 + 2. 6 + 1 + 2 = 9 so those three numbers can be crossed out. This leaves add 5 + 2 which gives us 7. Place the 7 on the left side of the x. Since the numbers on the left and right of the x are equal, the answer is correct.

Index

2-digit number, 118, 139
Activities, 7, 10, 14, 17, 19, 24, 26, 28, 30, 31, 33, 35, 36, 37, 42, 45, 49, 55, 78, 81, 82, 83, 91, 97, 98, 99, 101, 102, 103, 105, 106, 107, 108, 115, 117, 120, 129, 130, 131, 132, 134, 135, 138, 142, 143, 146, 150, 156, 174, 175, 177, 179, 181, 184, 191
Acute triangle, 85, 88
Addend, 135
Addition, 12, 15, 43, 46, 97, 112, 115, 126, 127, 129, 131, 132, 134, 135, 136, 137, 138, 141, 142, 145, 147, 151, 154, 155, 156, 158, 159, 161, 162, 176, 177, 181
Algebra, 42, 132, 151
Angles, 45, 78, 79, 80, 81, 82, 83, 85, 86, 87, 91, 98
Appropriate language, 11, 127
Area, 1, 13, 37, 47, 53, 78, 83, 84, 85, 86, 88, 89, 90, 91, 97, 98, 99, 101, 102, 103
Arrays, 146
Assessment, 3, 17, 22, 23, 33, 35, 37, 38, 39, 40, 41, 42, 43, 44, 55, 124
Attitude, 1, 13, 14, 17, 18, 19, 34
Attributes, 127
Basic concepts, 6, 14, 15, 43
Basic facts, 3, 14, 15, 25, 27, 41, 126, 127, 128, 129, 137, 141, 143, 151, 154, 159, 161
Beans, 105, 114, 115, 117, 133, 134, 149, 150, 197, 198
Bilingual,, 123
Bingo, 118, 178, 189, 191
Bridge the gap, 15, 33, 35, 97, 116, 125, 156, 174
Bridging the gap, 28
Bundling, 113, 114
Calculator, 24, 76, 89, 186
Calendars, 97
California Framework, 6, 36, 53, 54, 151
Child development, 115
Chinese, 45, 50, 82, 91
Circle, 85, 89
Circumference, 89, 90

Classification, 14
Classifying, 11, 77, 79, 129
Classroom, 3, 8, 9, 10, 12, 13, 16, 18, 20, 22, 23, 24, 25, 30, 32, 37, 38, 39, 40, 41, 42, 43, 44, 45, 49, 50, 52, 53, 54, 55, 56, 76, 79, 80, 82, 83, 93, 100, 101, 103, 104, 105, 108, 112, 113, 130, 131, 141, 158
Clock time, 105
Colored Chips, 156
Communicate mathematics, 10, 23
Commutative properties, 136
Comparing,, 11, 77, 180
Comparison 42, 80, 98, 99, 100, 101, 103, 104, 106, 107, 123
Composite numbers, 184
Computers, 8, 30, 76
Concept, 3, 5, 8, 11, 12, 16, 19, 20, 24, 25, 27, 28, 29, 30, 31, 33, 34, 35, 36, 37, 38, 40, 41, 46, 48, 51, 56, 76, 77, 78, 98, 101, 103, 104, 105, 113, 114, 115, 120, 124, 125, 126, 127, 129, 130, 131, 132, 135, 137, 142, 143, 146, 151, 154, 166, 171, 172, 177, 179, 184, 191, 192, 198
Cconcrete materials, 9, 15, 127
Concrete Operational, 27
Conservers, 128
Cuisenaire Rods, 132, 143
Cultural, 9, 16, 17, 19, 22. 46
Curriculum, 4, 5, 7, 8, 10, 12, 13, 14, 17, 18, 22, 23, 35, 36, 39, 41, 44, 47, 55, 77, 97, 169
Decimal, 46, 122, 124, 125, 126, 170, 190, 191, 192, 193, 198
Denominators, 180
Developmental stages, 77
Developmentally appropriate curriculum, 5, 10
Dewey's, 6
Diagnostic teaching, 24
Diameter, 89, 127
Digit cards, 116, 117, 120, 121, 122, 123, 124, 126, 135, 158, 159, 161, 174
Direct instruction, 35, 131, 132, 161
Disequilibrium", 25
Distributive method, 164, 165
Diverse needs, 18

Division, 46, 49, 50, 141, 142, 143, 145, 147, 149, 150, 151, 154, 162, 166, 167, 178, 183, 193, 197
Drill, 127, 130
English measurement, 97, 101
Environment, 9, 13, 19, 22, 25, 49, 76, 78, 119
Equal, 129, 145
Equity, 22
Error., 24
Euclidean concepts, 78
Evaluating, 11, 32
Expanded notation, 118, 164, 165
Factor trees, 185
Failure, 21, 48
Formulas, 37, 76, 77, 91, 97, 103
Fraction strips, 171, 180, 181, 182, 184
Fractions, 169, 170, 171, 172, 174, 175, 176, 178, 179, 183, 188, 191, 198
Free play, 28, 35, 79, 83, 129, 145
Games, 30, 177, 184
Geoblocks, 36, 38
Geometry, 17, 36, 42, 76, 77, 78, 79, 81, 82, 86, 91, 93, 97, 98, 129
Hands-on experiences, 10, 15, 28, 77, 115
Hindu – Arabic, 45
Homework, 15, 16, 42, 47, 48, 49, 50, 80, 104, 123
Hundreds, 112, 114, 121, 122, 123, 125, 156, 157, 159, 160, 163, 193
Identity element, 136
Internalize, 15, 16, 36, 112, 137
Internet, 30
Knowing, 8, 20, 22, 175
Language, 9, 10, 11, 12, 13, 15, 16, 17, 19, 26, 28, 32, 35, 36, 37, 43, 45, 48, 49, 50, 51, 77, 78, 82, 91, 112, 113, 123, 131, 132, 134, 146, 178, 185
Lattice multiplication, 165
Learning, 1, 2, 4, 9, 10, 11, 12, 13, 15, 17, 18, 20, 21, 22, 23, 25, 27, 30, 32, 33, 35, 39, 41, 42, 48, 49, 50, 51, 53, 54, 56, 77, 78, 79, 82, 83, 97, 108, 112, 115, 118, 119, 127, 128, 129, 130, 134, 137, 141, 142, 143, 145, 147, 148, 162, 169, 170
Learning Center Activities, 146
Lesson, 5, 13, 20, 21, 25, 31, 32, 33, 34, 35, 36, 37, 38, 52, 82, 171, 172, 174
Lesson plans, 18, 34
Linear measurement, 99, 106

listening to students', 9
Literature, 13, 14, 32, 46, 134, 137
Literature books, 13, 34, 136
Maintenance, 20, 34, 35
Manipulatives, 4, 6, 10, 13, 15, 16, 17, 20, 28, 29, 32, 33, 34, 35, 36, 49, 102, 112, 113, 114, 116, 117, 118, 120, 129, 130, 132, 135, 139, 140, 141, 143, 144, 145, 146, 149, 150, 151, 154, 155, 157, 158, 161, 162, 170, 171, 175, 176, 179, 183
Mass measurement, 107
Mathematical power, 41
Mathematics, 1, 2, 4, 5, 6, 7, 8, 10, 13, 14, 17, 20, 21, 22, 23, 31, 35, 40, 41, 42, 45, 46, 52, 53, 54, 81, 117, 129, 134, 143
Measurements, 85, 87, 90, 97, 99, 101, 102, 103, 105, 106, 107
Memorization, 3, 40, 49, 50, 53, 125, 126, 135, 137, 142, 143, 154, 169
Metric, 51, 97, 98, 101, 104
Money, 48, , 97, 99, 105, 106, 113, 190, 192
Motivation, 9, 32, 34, 37
Multidigit numbers, 115, 123, 151, 154, 161, 168
Multiples, 185
Multiplication, 50, 126, 137, 141, 142, 143, 145, 147, 148, 149, 150, 151, 154, 162, 163, 165, 166, 179, 182, 183, 185, 186, 187, 192
New new mathematics", 11
Non standard *measurement*, 99, 102, 103, 104, 106, 107
Non-conservers, 128
Non-proportional, 114
Number families, 135
Number operations, 97, 151
Number sentences, 132
Numeration blocks, 114, 155
Objective, 32, 34, 37, 40
Obtuse triangle, 85, 88
Palindromes:, 162
Paper airplanes, 107
Parallelogram, 78, 79, 80
Pattern blocks, 79, 80
Patterns, 14, 24, 32, 43, 76, 79, 80, 81, 82, 112, 129, 139, 140, 148, 154
Percentage, 170
Perimeter, 84, 90
Pi, 89
Piaget, 4, 5, 14, 25, 26, 27, 32, 78, 104
Pictorial level, 192

Place value, 112, 126
Planning, 20, 38
Portfolio, 35, 42, 44
Practice, 4, 5, 14, 15, 17, 20, 29, 30, 31, 34, 35, 37, 53, 83, 123, 126, 127, 129, 135, 137, 139, 146, 149, 150, 154, 157, 161, 165, 167, 182
Precision mrasurement, 99, 100, 101, 102, 103, 104, 105, 107
Preoperational, 26, 128, 142, 170
Prerequisite skills, 15, 21, 32, 37, 129, 142, 162, 175
Prime numbers, 184
Principles and Standards for School Mathematics, 6, 21
Probability, 36, 54, 197
Problem solving skills, 6
Proofs, 76
Proportional model, 114, 115, 162
Punishment, 16
Question, 2, 9, 12, 20, 24, 26, 31, 32, 33, 34, 41, 43, 44, 47, 49, 89, 105, 142, 170, 172, 175, 182, 183
Questioning techniques, 32
Quilts, 81
Radius, 89
Ratio, 170, 191
Rational numbers, 112, 154, 169, 170
Readiness, 32, 34, 175
Reading, 13, 14, 122
Reciprocal, 184
Rectangles, 84
Rhombus, 84
Sequential, 9, 20, 35, 36, 38, 53, 77, 97, 98, 99, 113
Seriation (Ordering, 127
Sorting, 11, 79
Spatial, 76, 77
Spiral, 15, 35, 36, 97
Square, 82, 85
Standard, 97, 99, 100, 101, 102, 103, 104, 105, 106, 107
Straight line, 81, 82, 86, 87, 88, 173
Strands,, 36
Strategies, 127, 135, 145, 147
Subtraction, 12, 46, 112, 116, 127, 129, 131, 132, 134, 135, 136, 138, 141, 142, 151, 154, 158, 159, 161, 162, 176, 177, 181, 182
Sum, 135
Symbols, 11, 12, 16, 19, 45, 46, 131
Symmetry, 83, 92
Take away, 160, 181
Tangram, 82

Teachers, 1, 2, 3, 4, 5, 8, 12, 13, 16, 17, 18, 19, 20, 21, 23, 24, 25, 30, 34, 36, 39, 40, 42, 46, 48, 49, 50, 52, 53, 56, 76, 77, 83, 97, 113, 115, 124, 132, 135, 137, 139, 151
Teaching, 2, 5, 6, 7, 8, 10, 11, 12, 14, 15, 17, 18, 20, 21, 22, 23, 24, 30, 32, 33, 34, 36, 37, 38, 39, 40, 41, 46, 47, 48, 51, 52, 53, 54, 55, 56, 76, 99, 113, 127, 134, 135, 137, 141, 142, 161, 171
Technology, 8, 9, 23
Tens, 112, 113, 114, 115, 116, 117, 118, 121, 122, 123, 125, 143, 148, 149, 150, 155, 156, 157, 158, 159, 160, 161, 162, 164, 166, 167, 192, 193
Tessellation patterns, 80
Test, 2, 24, 31, 33, 39, 40, 41, 48, 55, 56, 77, 80, 98, 126, 143, 145
Testing, 11, 18, 30, 35, 39, 40, 55, 80
Textbook, 5, 6, 17, 18, 24, 30, 31, 33, 34, 35, 77, 112, 122, 124, 129, 132, 143, 161
Thinking strategies, 127
Time measurement, 104
Timed tests, 4, 17, 55, 56, 126, 143
Topological concepts, 78
Trapezoid, 79, 80, 82
Triangle, 85, 87
Trillions, 122, 124
Understanding, 3, 4, 5, 6, 7, 8, 10, 11, 13, 14, 15, 17, 19, 20, 22, 23, 27, 28, 30, 35, 37, 42, 48, 53, 55, 56, 77, 98, 103, 108, 112, 115, 126, 129, 135, 141, 154, 162, 167, 172, 192, 198
Unifix cubes, 104, 107, 117, 130, 132, 143, 145, 150
Upper grades S, 120
Visualization skills, 29
Vocabulary, 11, 17, 28, 29, 32, 35, 36, 37, 86, 98
Volume, 103
Walkabout assessment, 33, 38
Weight mesurementT, 103
Whole language, 11, 134
Whole numbers, 45, 124, 125, 154, 173, 183, 192
<u>*Worksheet*</u>*, 174, 181, 182, 187, 188, 190*
Writing, 3, 5, 9, 10, 12, 13, 23, 36, 37, 41, 42, 44 , 91, 115, 117, 132, 134, 136, 140, 150, 173, 174, 193

Bibliography

Amenis, Jerry A. and J.V. Ebenezer, (Series Editors). *Mathematics on the Internet, A Resource for K-12 Teachers.*. Upper Saddle River, New Jersey. Prentice Hall, 2000

Ashlock, R. Error *Patterns in Computation*. Columbus. Charles Merrill, 2001

Banks, JA and CA Banks. *Multicultural Education, Issues and Perspectives*. 3rd Edition. New York. John Wiley & Sons, Inc, 1999

Burns, M. *Multiplication*. Math Solutions Publications, 1991

California Deparment of Education. *Mathematics Framework for California Public Schools*. California Department of Education, 2000 Revision

California Math Council. *Assessment Alternatives in Mathematics. An overview of Assessment Techniques that Promote Learning*. Berkley. Lawrence Hall of Science, 1989

Coats, Grace and J. Stenmark. *Family Math for Young Children*. Berkley. Lawrence Hall of Science, 1997

Coerr, E. The Josefina Story Quilt. Harper and Row, 1986

Ernst, L. and L.C. Ernst. *The Tangram Magician*. Harry Abrams, Inc., 1990

Felter, S.A. *The Analysis of Written Arithmetic*. Book First. New York. Charles Scribner's Son, 1862

Goodman, J. *Group Solutions*. Berkley. Lawrence Hall of Science, 1992

Goodman,J. & J. Kopp. *Group Solutions, Too*. Berkley. Lawrence Hall of Science, 1997

Griffiths, R. and M. Clyne, M. *Language in the Mathematics Classroom*. Portsmouth. Heinmann, 1994

Gwynn, F. *Pondlarker*. New York. Simon & Schuster Books for Young Children, 1990

Jenkins, L, Laycock M and McLean Peggy. *Geoblocks and Geojackets*. Hayward CA. Activity Resources Company, Inc., 1988

Leedy, L. *2 x 2 = BOO!* New York. Holiday House, 1995

Lionni, L. *Swimmy*. New York. Scholastic, Inc., 1991

Lorton, M.B. *Mathematics Their Way*. Addison Wesley, 1994

Milne, W. J. *Standard Arithmetic*. New York. American Book Company, 1892

Myller, R. *How Big is a Foot*. New York. Batam Doubleday Dell Books for Young Readers, 1990.

National Council of Teachers of Mathematics. *Assessment Standards for School Mathematics*. Reston, VA. National Council of Teachers of Mathematics, 1995

National Council of Teachers of Mathematics, *Curriculum and Evaluations Standards for School Mathematics*. Reston VA. National Council of Teachers of Mathematics, 1989

National Council of Teachers of Mathematics. *Principles and Standards for School Mathematics*. Reston, VA. National Council of Teachers of Mathematics, 2000

National Council of Teachers of Mathematics. *Readings in the History of Mathematics Education*. Edited by James Bidwell & Robert G. Clason. National Council of Teachers of Mathematic. Washington D.C., 1970.

National Council of Teachers of Mathematics. *Developing Computational Skills*. 1978 Yearbook. Edited by Marily Suydam and Robert Reys. Reston, National Council of Teachers of Mathematics, 1978

Piaget, Jean. *The Child's Concept of Number*. New York. W.W. Norton & Co. 1965

Prince, J.T. *Arithmetic by Grades*. Boston. Ginn & Co., 1894

Ray, J. *Ray's Arithmetical Key. The Second Book*. Cincinnati. Wilson, Hinkle & Co, 1845

Shuard H. and Rothery A. *Children Reading Mathematics*. John Murray, 1984

Silverstein, S. *The Giving Tree*. Harpercollins Juvenile Books, 1986

Stenmark, JK, Thompson, V and Cossey R. *Family Math*. Berkley. Lawrence Hall of Science, 1986

Thompson, J.B. *Complete Graded Arithmetic*. Chicago. Clark & Maynard Publishers, 1884

Thompson, V. and Mayfield-Ingram, K. *Family Math – The Middle School Years*. Berkley. Lawrence Hall of Science, 1998

Tompert. A, *Grandfather Tang's Story*. New York. Crown Publishing Co, 1990

Van de Walle, J. *Elementary School Mathematics*. New York. Longman, 1990

Willoughby, S. *Mathematics Education for a Changing World*. ASCD, 1990

Wisconsin Research and Development Center for Cognitive Learning. *Developing Mathematical Processes*. Chicago. Rand McNally & Company, 1974

------ *Math Talk*. Portsmouth NH. Heinemann, 1987

------*Language in Mathematics*. Edited by Jennie Bickmore-Brand. Portsmouth. Heinemann, 1990